Artifacts of Political Power

Artifacts of Political Power

How Historical Objects Shaped Governments and Empires

Charles Kensington

Hidden Histories Media

Staunton, VA 24401

Printed in the United States of America

First Printing, 2025

Published by Hidden Histories Media

Paperback ISBN: 978-1-971137-17-9

Contents

Introduction vii

Chapter 1 The Clay That Commanded — The Cyrus Cylinder and Imperial Authority 13

Chapter 2 The Stone That Fixed the Law — Hammurabi's Stele and Public Authority 21

Chapter 3 The Decree in Two Tongues — The Rosetta Stone and Imperial Control 29

Chapter 4 The Eagle That Held the Line — Roman Military Standards and Imperial Command 37

Chapter 5 The Pillar That Carried the King's Voice — Ashoka's Edicts and Moral Rule 45

Chapter 6 The Jade That Spoke for Heaven — The Heirloom Seal of the Realm 53

Chapter 7 The Stone That Guarded the Fields — Kudurru Boundary Markers and Land Authority 61

Chapter 8 The Ring That Sealed the Crown — Egyptian Signet Rings and Royal Authority 69

Chapter 9 The Rock That Marked the Border — The Behistun Inscription and Imperial Claim 77

Chapter 10 The Shield That Set the Limits — The Twelve Tables and Roman Civil Authority 85

Chapter 11 The Spear That Chose a King — The Frankish Lance and Royal Acclamation 93

Chapter 12 The Crown That Bound the Realm — The Hungarian Holy Crown and Legal Kingship ...101
Chapter 13 The Drum That Summoned the Court — The Zhou Dynastic Bells and Ritual Authority......109
Chapter 14 The Banner That Defined a People — The Aztec Codex Mendoza and Imperial Tribute...117
Chapter 15 The Charter in Wax and Silk — The Magna Carta and Constrained Kingship.........125
Chapter 16 The Codex That Divided a Faith — The Donation of Constantine and Papal Authority.. 133
Chapter 17 The Tablet That Bound the People — The Aztec Tribute Stone and Imperial Obligation .141
Chapter 18 The Sword That Crowned the Realm — The Japanese Imperial Regalia and Sacred Rule ..150
Chapter 19 The Tablet That Rewrote a Kingdom — The Sumerian King List and Political Memory....159
Chapter 20 Epilogue: The Object That Outlived the Throne — Power Preserved in Material Form 168

Introduction

In the dim light of a museum gallery, a single spotlight falls on a small clay cylinder. Its surface is worn, its edges chipped from centuries of handling, yet the wedge shaped marks remain clear. Each line of cuneiform catches the faint glow, drawing the eye into a story almost imperceptible at first glance. A security guard paces silently behind the glass, his footsteps echoing against polished stone floors. Outside, the city hums with the ordinary sounds of traffic and conversation, but inside, time bends. In this quiet room, a Persian king's voice still speaks, pressed into clay, commanding authority, announcing laws, and shaping the lives of thousands he would never meet. One cannot touch it, yet it exerts influence across the centuries, an object that holds the weight of an empire in its fragile form.

How does something so small, so inanimate, become a vessel for power? How can a carved stone, a polished crown, a mirror, or a jewel enforce rule where words alone cannot? These are not symbols of authority in a vague or artistic sense. They are working instruments, engineered, managed, and protected to control the mechanisms of governance. From the kings of Mesopotamia to the emperors of Japan, rulers discovered that the right object, held in the right way, could bend institutions, shape loyalties, and redefine succession. The question arises: how do material things, made by human hands, reshape human behavior on a scale that exceeds the immediate moment of creation?

The thrill of discovery begins with touch, even if only imagined. Historians, archaeologists, and curators have long approached these objects with awe, sometimes fear. They open doors to rooms sealed for centuries, uncover manuscripts hidden in chests or vaults, and brush away dust from inscriptions that survived fire, flood, and war. Each artifact contains a story of human intent, design, and consequence, yet the story is never straightforward. A seal once pressed into clay could authorize a decree that toppled officials. A crown resting on a brow could legitimize an entire dynasty. A stone inscription might survive long after a ruler's life, creating authority in absentia. The awe comes from realizing that human power is not only declared. It is manufactured, mediated, and maintained through material forms that endure long beyond the fragile human actors who created them.

This book traces that thread, from the first impressions pressed into clay in the cities of Sumer, through the carefully curated regalia of emperors, to codices and carved tablets that chronicled tribute, law, and obligation across continents. Readers will move through sunlit courts and shadowed archives, into the hands of scribes and priests, through ceremonies designed to awe, enforce, and preserve. The journey encompasses Egypt, Mesopotamia, Persia, China, Japan, Mesoamerica, and medieval Europe, each scene anchored by the objects that carried authority across space and time. You will witness bronze bells tuned to affirm rank, jade seals pressed to confirm rule under heaven, and wax impressed charters that constrained kings. These are not stories of symbols alone. They are tales of function, maintenance, secrecy, and the human impulse to control knowledge and power through matter.

Some objects were lost, hidden, or captured, yet even in absence they influenced decisions. Boundary stones protected legal claims long after kings died. Manuscripts were locked away in monastic libraries, surviving neglect and fire. Crowns were secreted in treasuries, accessible to only a chosen few, their absence lending tension and authority at the same time. Through these episodes, a pattern emerges: institutions do not destroy when knowledge or power is dangerous. They contain, restrict, and regulate. Silence, omission, and controlled access are as potent as any seal or sword. By studying these artifacts and the decisions surrounding them, one begins to see a continuum. The measures used to control authority in the past resonate in archives, museums, and vaults today.

The emotional core of this story lies in proximity. Stand before a stone stele, a sealed clay cylinder, or a crown behind glass, and one senses the palpable tension between human ambition and material endurance. The awe comes from the realization that power was literally imprinted into matter, preserved for centuries, and in some cases, still capable of shaping perception, ceremony, and legitimacy. Scholars have noted that handling, copying, or even reading these objects could provoke hesitation, fear, or extraordinary care. Institutions, aware of the potential consequences, developed protocols that survive in documentation, showing a human impulse to preserve what is dangerous while still wielding its authority.

Readers will encounter patterns repeated across cultures: knowledge deemed too risky to share freely is confined. Objects that enforce power are guarded rigorously. Texts and inscriptions are copied, annotated, and hidden. Ceremonial or administrative practices extend influence long after the original actors have vanished. A seal might authorize an edict that travels across provinces. A crown or sword confirms succession in ways that military strength alone cannot. Each example illustrates a fundamental principle: material artifacts do not merely reflect authority. They create it. Their endurance demonstrates how human institutions have long grappled with the tension between dissemination and control, knowledge and fear.

This is the journey of the book: to follow those objects, to witness how they were created, preserved, and deployed, and to understand the human behaviors surrounding them. You will see decisions made in haste and in fear, compromises between access and secrecy, and the long arc of consequences that follow every imprint pressed into clay, every blade polished for ritual, every stone carved with law. These objects endured through wars, neglect, and environmental decay, yet they remain legible to those who know how to read them. Their survival is a testament to the foresight, and sometimes the caution, of those who understood that power is most effective when it is tangible and controlled.

By the end of this book, you will have traced authority across continents and centuries, observing the repeated strategies humans have employed to enforce rule through matter. The stakes are both practical and symbolic. When a tablet records succession, when a crown confirms legitimacy, when a charter constrains a king, each act involves human judgment, ritual, and careful handling. These are not relics of myth. They are instruments of governance that shaped lives, cities, and empires.

As you turn the pages, consider the objects themselves: their weight, their texture, their precise inscriptions, and the countless hands that managed them over centuries. Each carries a story of enforcement, of care, and of institutional restraint. From clay cylinders to jade seals, from bells to crowns, they embody a human understanding that power and knowledge must be mediated through physical form, and that restraint is sometimes the greatest tool of authority.

The opening scene in the museum gallery, where a single cylinder catches the light, mirrors the arc of the book. Small, seemingly inert, yet capable of commanding attention, shaping decisions, and persisting beyond any single lifetime. Each artifact in these pages holds similar potential, and each reveals the complex interplay between human intention, material durability, and institutional care. The journey ahead is both historical and immediate, inviting the reader to witness the ways in which authority has been preserved in objects and the quiet but enduring tension between knowledge, power, and control.

Step now into the courts, temples, archives, and plains where these objects once functioned. Stand alongside scribes, priests, officials, and attendants who managed them. Watch the tension as a crown is lifted, a seal pressed, a tablet set in place. Listen to the silence that follows each ceremonial act, where authority is confirmed not by speech alone, but by the careful, deliberate presence of matter that endures. The story begins here, with each object a testament to human ingenuity, fear, and foresight, a story that will unfold in the chapters to follow.

Chapter 1
The Clay That Commanded — The Cyrus Cylinder and Imperial Authority

Before sunrise in 539 BCE, the air over Babylon hangs heavy with smoke from cooking fires and the dust of trampled streets. The city's vast walls still rise above the river plain, but its rule has shifted. Persian soldiers move through the gates in ordered lines, their armor catching the first pale light. Inside the Esagila temple complex, scribes gather in a quiet chamber. Reed mats cover the floor. Trays of damp clay sit nearby, cool and dark. At the center rests a smooth clay cylinder, newly formed and waiting.

A senior scribe steadies the cylinder with one hand and lifts a stylus with the other. He presses wedge shaped marks into the soft surface, line by careful line, curving the text around the clay. The words speak in the name of Cyrus, king of Persia, now ruler of

Babylon. Outside, priests, merchants, and officials adjust to a new sovereign. Inside this room, authority takes physical form. Each mark fixes a claim of power that will soon be set, hardened, and placed where it can be found.

A City Strained by Absence

For years before Cyrus entered Babylon, the city had lived with uneasy silence at its center. King Nabonidus had spent long stretches away in the Arabian oasis of Tayma. In his absence, temple rituals that honored Marduk faltered. Priests recorded missed festivals and altered ceremonies. The great New Year rites, which confirmed royal duty before the god, were disrupted. In a city built on sacred order, that disruption carried weight.

Administrators continued their work. Tax records were pressed into clay. Grain shipments moved along canals. Yet officials wrote under growing tension. Some tablets note complaints about temple neglect and shortages in offerings. Merchants and priests watched the throne with concern. Authority still existed, but its daily presence felt thin. In this climate, a new ruler would need more than soldiers at the gates. He would need the trust of Babylon's institutions.

Words Pressed Into Wet Clay

In the days after Cyrus entered Babylon, temple scribes prepared a cylinder of fine clay taken from local riverbeds. The clay was cleaned of grit and shaped by hand into a smooth barrel form. While still damp, it was set on a low surface and turned slowly as a scribe pressed lines of Akkadian cuneiform into the surface. The script followed established temple style, using the formal language of royal proclamations known in Mesopotamia for centuries.

The text was not improvised. It drew on earlier models of foundation inscriptions placed inside walls or beneath temples. Scribes listed Cyrus's titles, his lineage, and his claim to have been chosen by Marduk. Once inscribed, the cylinder was dried and baked to harden the clay. It was then placed within the rebuilt temple structure, where it became part of the building itself.

Quiet Murmurs in the Temple Courts

Word of the new inscription moved first through temple corridors. Priests who oversaw the rebuilding of Esagila were aware that a cylinder had been placed within the structure. They knew the language it carried, even if most citizens would never see it. Senior clergy read the lines that praised Marduk and blamed Nabonidus for neglect. Some nodded at the return of proper ritual order. Others read more slowly, aware that the words reshaped recent memory.

Outside the temple, city officials heard reports that Cyrus had restored images of gods to their shrines and allowed displaced communities to return home. Merchants and local elites watched carefully. The conquest had been swift and mostly peaceful, yet the presence of foreign troops remained visible. The inscription promised stability, but promises in clay could not erase uncertainty overnight. In private, some wondered how firmly this new king understood Babylon's traditions.

Watchful Eyes Among Babylon's Leaders

As weeks passed, senior temple officials met in shaded courtyards to speak in guarded tones. Cyrus had honored Marduk in the inscription, yet he remained a foreign ruler with his own traditions and court advisers. Priests who managed temple estates worried about tax obligations and control over sacred lands. If Persian administrators altered long standing privileges, temple income could shift. The cylinder's words promised respect, but officials knew that policy was carried out in daily orders, not carved claims.

City governors and scribal supervisors also felt pressure. They were responsible for keeping grain supplies steady and records accurate under the new regime. Persian envoys began appearing in administrative offices, observing procedures and reviewing accounts. Some Babylonian officials feared replacement if they failed to adapt. Others hesitated to act too quickly, uncertain how much authority still rested in local hands. Public calm masked private tension as each group tested the limits of Cyrus's rule.

Guarded Words Beneath the Temple Walls

Temple scribes did not treat the cylinder as a public notice. Like earlier foundation texts, it was placed within the structure of the rebuilt temple, not displayed in an open square. Access was limited to those overseeing construction and ritual spaces. The wording followed long standing royal formulas, which trained scribes could read with ease. In administrative tablets from the period, officials continued to record offerings and repairs, showing that the inscription aligned with ongoing temple work.

There is no record that crowds gathered to study the text. Later writers would sometimes describe it as a broad declaration of rights, but no ancient source shows public reading ceremonies in Babylon. The evidence points instead to controlled placement within sacred architecture. By embedding the text in the temple foundation, Cyrus fixed his claims in a setting that temple authorities monitored closely. Control rested not in spectacle, but in restricted access.

Careers Balanced on a New King's Favor

For Babylonian officials who had served Nabonidus, the shift in power carried personal risk. Some had signed tablets in the former king's name for years. Their seals marked tax orders, land grants, and temple accounts. Now those same hands pressed impressions under Persian oversight. A mistake in loyalty could bring dismissal or loss of status. Surviving records show that many administrators continued in office, yet the change required careful public alignment with Cyrus's rule.

Temple leaders also faced strain. If they openly resisted Persian authority, they risked losing support for repairs and offerings. If they embraced the new king too quickly, they could damage their standing among traditionalists who still felt loyalty to the past. Families tied to temple service watched closely. Reputation rested on appearing steady and devout. In private spaces, conversations grew cautious. The clay cylinder hardened in its place, while those around it adjusted their words and alliances.

Orders Carried Through Official Channels

In the months after the conquest, administrative tablets show steady activity under Cyrus's name. Tax collections continued. Temple rations were measured and distributed. Local officials dated documents by the regnal year of the new king. These changes followed established procedure. Scribes updated headings on clay tablets, and seal impressions shifted to reflect Persian authority. The process was gradual but firm, carried out through routine record keeping.

Within the temple complex, rebuilding efforts moved forward under royal support. Records note materials delivered and workers assigned to repairs. The cylinder itself remained embedded in the structure, not moved or copied in large numbers. No decree ordered its removal or alteration. Instead, the text functioned as a fixed foundation statement while daily governance adapted around it. Persian administrators observed, approved, and at times confirmed local decisions through written orders that reinforced the new chain of command.

Echoes Through Later Reigns

Years after Cyrus's death, Babylon remained part of the Persian Empire. Administrative tablets continued to date events by the reigns of Cambyses and Darius. The temple of Marduk still stood, and the cylinder lay within its walls. The language used in later royal inscriptions across the empire followed similar patterns, presenting each ruler as chosen by a chief god and protector of local worship. The formula proved durable.

Yet loyalty in Babylon did not settle into quiet certainty. Rebellions broke out during the reign of Darius I, and some insurgents claimed royal titles rooted in older traditions. Persian forces suppressed these uprisings, and new inscriptions recorded victory. The earlier promise of restoration did not prevent unrest. The cylinder's words endured in clay, but the balance between imperial control and local trust required constant renewal, generation after generation.

Buried Cylinder, Unearthed Claims

In 1879, during excavations at Babylon led by Hormuzd Rassam on behalf of the British Museum, workers uncovered fragments of a baked clay cylinder within the ruins of the Esagila complex. The pieces were later joined and studied in London. The cuneiform text matched the style of Neo Babylonian inscriptions and confirmed Cyrus's titles and claims. Scholars compared it with other foundation texts and with administrative tablets from the same period.

The cylinder did not emerge whole. Parts were damaged, and some lines required reconstruction from parallel texts. Over time, additional fragments were identified and joined. In the twentieth century, the object was displayed as evidence of Persian policy in Babylon. Modern study relies on the physical clay, the script, and related tablets. The surface still bears the pressure of the stylus. The words remain fixed where a scribe once pressed them into wet clay.

Chapter 2
The Stone That Fixed the Law — Hammurabi's Stele and Public Authority

Under the sharp sun of eighteenth century BCE Babylon, laborers brace their shoulders against thick ropes as a dark basalt stele rises inch by inch. Wooden rollers grind against packed earth. Sweat runs down bare backs. The stone is heavy, cut from hard rock brought from far away. Its surface has been smoothed and carved with care. At the top, a relief shows King Hammurabi standing before Shamash, the sun god, who sits on a throne and extends a rod and ring. Below the image, lines of cuneiform descend in tight columns.

As the monument settles upright, officials step forward to inspect it. Scribes trace the carved signs with their eyes. A crowd gathers at a distance, murmuring in the heat. The words are no longer confined to clay tablets in palace rooms. They are cut into stone for all to see. Dust drifts through the air as the ropes fall slack and the stele stands firm in the open light.

A Kingdom Forged by Campaign and Control

When Hammurabi took the throne around 1792 BCE, Babylon was not yet the dominant power in southern Mesopotamia. Rival city states controlled nearby rivers and trade routes. Alliances shifted with the seasons. Armies marched across canal fed fields, and fortified walls marked fragile borders. Hammurabi spent years securing his position through treaties, marriages, and military campaigns. Gradually, cities that once resisted Babylon's rule fell under his command.

As his territory expanded, so did the need for consistent judgment. Local governors reported disputes over land, debts, and labor. Temple courts handled cases tied to offerings and contracts. Palace officials kept records on clay tablets that tracked fines and penalties. Authority depended on written decisions that could be enforced beyond the capital. In this growing kingdom, law had to travel across distance and remain stable under many hands.

Carved Words in Black Basalt

The stele was cut from a block of dark basalt, a stone harder than the clay tablets used in daily record keeping. Basalt did not come from the plains around Babylon. It had to be quarried and transported, likely from distant regions where such stone could be found. Skilled stoneworkers shaped the tall slab into a smooth pillar, tapering slightly toward the top. The surface was polished before any carving began.

At the upper section, artisans carved a relief of Hammurabi standing before Shamash, the god associated with justice and the sun. Below this image, trained scribes drafted the legal text in cuneiform signs. Stone carvers then chiseled the characters into the surface with precision. The text was long and tightly arranged, requiring steady hands and careful planning. Each line was fixed in stone, not pressed into clay that could be reshaped.

Murmurs Before the Monument

When the stele was set in place, palace officials were the first to examine it closely. Senior scribes read through the carved laws, checking familiar phrases and penalties. Judges who had long relied on written tablets now faced a public monument that carried the king's voice in permanent form. Some nodded at the clarity of fixed standards. Others studied the details, aware that the stone reduced room for quiet adjustment in private hearings.

Merchants, craftsmen, and landholders gathered in the square, listening as sections were read aloud. Many heard rules that touched daily life, from contracts and wages to injury and debt. The penalties were clear and often severe. Some in the crowd shifted uneasily. The monument stood beyond argument, yet daily disputes still required human judgment. As the crowd thinned, conversations continued in doorways and workshops, weighing how these carved lines would shape future cases.

Judges Under the King's Shadow

As news of the stele spread across Babylon and its subject cities, local judges felt new pressure. They had long handled disputes with reference to custom, prior rulings, and written contracts. Now a tall monument stood in the capital, declaring fixed penalties for theft, injury, false accusation, and failure in duty. Litigants began to refer to the king's carved words. A judge who ruled too far from the stele's standards risked complaint.

Temple courts also took notice. Priests oversaw property tied to sacred estates and enforced vows made before the gods. The relief at the top of the stele showed Hammurabi before Shamash, linking royal law with divine authority. Some religious officials accepted this pairing. Others watched carefully, aware that public law might limit their discretion. In private rooms, conversations grew cautious as officials measured how closely they must align each decision with the stone in the square.

Reading the Stone, Guarding Its Meaning

Scribes began copying sections of the stele onto clay tablets for training and reference. In schoolrooms attached to temples and palaces, students practiced the legal formulas carved on the monument. Surviving tablets from later periods show similar phrasing, suggesting that the text was studied as a model of royal justice. These copies were not exact replicas of the full stele, but they carried selected laws into daily practice.

There is no record of guards stationed beside the monument at all hours, yet its location in a central public space placed it under watch. Officials controlled formal readings and legal citations. Later tradition sometimes described the stele as a complete code applied in every case. Surviving court tablets, however, show that judges still weighed evidence and circumstance. The stone fixed the king's words, but their use required trained interpretation.

Lives Measured Against Carved Penalties

For judges in Babylon's courts, each ruling now carried sharper risk. If a verdict strayed too far from the stele's penalties, a losing party could claim injustice before higher officials. A judge accused of favoritism or error faced fines and removal from office under certain provisions. The monument did not simply guide decisions. It placed the king's authority over every courtroom. Some officials grew cautious, limiting their discretion to avoid scrutiny.

For ordinary citizens, the cost could be severe. Builders, physicians, and ship captains saw their work tied to strict consequences. A collapsed house or failed operation could bring heavy punishment. Families felt strain when debts or injuries were judged under fixed standards. Reputation mattered deeply in tight communities. A public case could stain a household for years. As disputes rose before judges, the stone's carved lines loomed over personal fate and daily survival.

Law Carried Across a Growing Realm

As Hammurabi's rule extended over former rival cities, officials worked to align local courts with Babylonian standards. Governors sent reports to the capital that recorded disputes and penalties. Clay tablets from distant towns began to reflect similar legal language. Royal correspondence instructed administrators to uphold justice in the king's name. The stele in Babylon stood as a fixed model while copies and summaries traveled through trained scribes.

Within the palace, records were archived and stored in ordered rooms. Decisions were dated by the king's regnal years, tying legal action to his authority. When new territories were absorbed, local customs were not erased at once, but they were framed within royal oversight. The monument itself remained in Babylon, not moved to each province. Its authority moved instead through written orders and trained officials who carried its standards into daily governance.

A Monument Taken and Remembered

Decades after Hammurabi's death, Babylon's power shifted. Armies from Elam entered the region and carried off treasures from the city. Among the captured objects was the basalt stele. It was transported east and set up in the Elamite capital of Susa. There, the monument stood in a foreign land, removed from the courts it once overshadowed. Its lower section was damaged, likely to make space for a new inscription that was never completed.

In Babylon, law continued through tablets and local rulings, yet the physical monument was gone. The king's carved image no longer faced the city square. Officials relied on archived texts and trained memory. The absence of the stele did not erase the legal system, but it altered the visible anchor of authority. For generations, the stone remained far from the place where it first fixed the king's words.

Unearthed in Susa's Ruins

In 1901, French archaeologists excavating the ancient city of Susa uncovered fragments of a tall basalt stele buried among the ruins. The pieces were lifted carefully from the soil and later assembled. The relief of Hammurabi and Shamash remained visible, along with most of the long legal inscription. Damage at the bottom showed where the stone had been altered after its removal from Babylon. The monument was transported to Paris, where it was studied and displayed.

Scholars compared the carved laws with thousands of clay tablets found in Mesopotamian sites. The language matched known legal formulas of the period. The stone's weight, polish, and tool marks confirmed its careful creation. Visitors today stand before it behind glass, reading translations printed nearby. The basalt still carries the marks of the chisel. The king and the god face one another in silent relief, as the carved lines descend toward the damaged edge.

Chapter 3
The Decree in Two Tongues — The Rosetta Stone and Imperial Control

In 196 BCE, under the heavy sky of the Nile Delta, stoneworkers brace their feet against packed earth as chisels strike a dark granodiorite slab. The sharp ring of metal carries across the temple courtyard. Dust gathers in the folds of linen robes. Nearby, priests in white garments review a draft of the decree, their fingers tracing wax tablets before the words are cut into stone. Greek officials stand close, watching each line take shape.

The slab is tall and thick, carved to endure flood and sun. At the top, Greek text spreads in clear, formal lines. Beneath it, Egyptian scripts flow in tighter bands, one used by priests, the other in everyday writing. The same royal decree appears three times, fixed into one surface. Laborers pause to wipe grit from the carved letters. When the slab is raised upright, it will stand as a public record of loyalty, reward, and command.

A Kingdom Ruled by Two Traditions

After the death of Alexander the Great, Egypt passed into the hands of one of his generals, Ptolemy. His family ruled as Greek speaking monarchs over a land with deep Egyptian traditions. By the time Ptolemy V took the throne, he was still a child. Court advisers and regents governed in his name. Rebellions had broken out in parts of the country, and rival claimants threatened stability along the Nile.

Temples held strong influence across Egypt. Priests controlled land, wealth, and local loyalty. Royal authority depended on their support. Greek administrators managed taxes and trade, while Egyptian religious leaders guarded ritual order. In cities like Memphis, these systems overlapped. Decrees had to satisfy both the Greek court and the priesthood. In this divided structure of power, a written proclamation carried weight only if both traditions recognized it.

Carving a Royal Order in Three Scripts

The decree that would be cut into the stone was first agreed upon at a gathering of priests in Memphis. Records from later copies show that the priests drafted the text to praise Ptolemy V and confirm benefits granted to temples. The wording followed formal patterns used in earlier Egyptian decrees. Once approved, the text was prepared in three versions: Greek for the royal court, Demotic for daily administrative use, and hieroglyphic for sacred display.

Stoneworkers selected a block of granodiorite, a hard, dark stone that could endure weather and time. The surface was smoothed, then divided into sections. Skilled carvers worked from prepared drafts, cutting the Greek text at the top and the Egyptian scripts below. Each line required steady pressure and careful spacing. The finished slab was heavy and thick, meant to stand upright in a temple setting where officials and priests could oversee it.

A Proclamation Read Before the Gods

When the stone was set upright in the temple courtyard, priests gathered first to hear the decree read aloud. The hieroglyphic text affirmed royal gifts, tax reductions, and honors for the temples. Some nodded at the promises of restored order and protection. Younger priests listened closely, aware that support for the boy king carried risk if rival forces gained strength elsewhere along the Nile.

Greek officials and local administrators also took note. The Greek version confirmed loyalty to Ptolemy V and condemned rebellion. Merchants and town leaders heard that certain debts were remitted and that temples would receive privileges. Yet beyond the formal praise, questions lingered. Armed conflict had not vanished from Egypt. The decree stood firm in stone, but the young king's authority still depended on distant armies and fragile alliances.

Priests and Regents Under Watchful Eyes

As copies of the decree were ordered for other temples, pressure grew among both priestly councils and royal regents. Priests who endorsed the text tied their public loyalty to a young king whose hold on Upper Egypt was not secure. Some temple leaders worried that if rebels advanced, their open support for Ptolemy V could bring retaliation. Endorsing the decree was not a silent act. It was carved in stone for all to see.

At the royal court, advisers felt equal strain. The decree promised rewards and tax relief, which affected state revenue. Greek administrators had to balance temple privileges with military needs. Officials exchanged reports about unrest in certain regions. Each new inscription fixed the king's claims more firmly, yet it also widened the circle of those publicly committed to his rule. Silence became harder to maintain as the stones multiplied across the land.

Reading Across Scripts, Guarding Authority

Temple scribes were responsible for preserving and copying the decree. Surviving records show that multiple stelae bearing the same text were erected in different temple centers. Priests ensured that the hieroglyphic and Demotic versions matched the approved wording. Greek officials reviewed the Greek text for accuracy. These checks were part of routine temple and court procedure, not informal efforts. The text was controlled through authorized copies rather than open revision.

There is no ancient record of disputes over the wording carved on the Rosetta Stone itself. Later writers sometimes described the decree as a broad charter of rights, but surviving sources present it as a formal priestly resolution honoring the king. Access to the stone rested within temple grounds. Public reading took place under priestly supervision. The scripts required trained readers, limiting who could fully grasp the full inscription without assistance.

Loyalty Carved Into Reputation

For senior priests who endorsed the decree, their names and offices were recorded in temple archives tied to the resolution. Public support for Ptolemy V strengthened their standing with the royal court, yet it also placed them in clear view of critics. If rebellion succeeded in any region, their open alignment with the young king could bring loss of position or worse. Families linked to temple offices felt the weight of these decisions.

Greek administrators faced strain as well. They were responsible for enforcing tax adjustments promised in the decree while still meeting military expenses. A miscalculation could draw blame from both the palace and local communities. Some officials were transferred or replaced during periods of unrest. In towns along the Nile, loyalty became a visible marker of safety or risk. The stone's carved praise bound individuals to a ruler whose future was not fully secure.

Copies Raised Across the Nile

After the decree was approved, temple authorities arranged for additional copies to be carved and set up in major religious centers. Surviving fragments from sites beyond Rosetta confirm that the text was reproduced under official direction. Each stone followed the same structure, with Greek and Egyptian scripts arranged in ordered bands. Local priests supervised placement within temple courtyards, ensuring the decree stood in visible, sanctioned spaces.

Administrative offices updated tax registers and temple accounts in line with the promises recorded in the decree. Orders moved through written instructions carried by couriers along the river. Greek officials monitored compliance, while temple scribes recorded offerings and privileges granted. The stones themselves were not moved once installed. They remained fixed in sacred precincts, where access was shaped by temple custom and daily ritual practice rather than open civic display.

A Kingdom That Did Not Stand Still

Years after the decree was carved, Ptolemy V grew into adulthood and continued to rule, yet unrest did not fully disappear. Revolts flared in parts of Upper Egypt, and royal forces were sent to restore control. The promises set in stone did not prevent shifting alliances or local resistance. Temples remained centers of influence, but their loyalty required steady negotiation with the crown.

As decades passed, new decrees were issued under later rulers, each shaped by fresh political pressures. The Rosetta decree became one record among many carved across the Ptolemaic realm. Some stones were damaged by weather or later construction. Others stood in courtyards where regimes changed around them. The authority fixed in three scripts endured on stone, even as the balance between Greek rulers and Egyptian traditions remained unsettled.

A Fragment Found in Fortified Walls

In 1799, French soldiers strengthening a fort near the town of Rosetta uncovered a dark stone built into the wall. The slab was broken, its upper portion missing, yet three bands of text were clearly visible. Officers recognized the scripts as Greek and Egyptian. The stone was removed from the masonry and examined more closely. Scholars quickly copied the inscriptions before the artifact changed hands during the conflict that followed.

After the British took control of the region, the stone was transported to London, where it entered the British Museum. Its surface still bore the marks of ancient chisels and later damage. Researchers compared the Greek text with the Egyptian scripts, using the parallel wording to unlock long forgotten signs. The broken slab stood under glass, its carved lines linking a royal decree to centuries of silence.

Chapter 4
The Eagle That Held the Line — Roman Military Standards and Imperial Command

At first light on a cold plain in Germania, Roman soldiers stand in tight ranks, shields locked and boots pressed into damp ground. Mist drifts low across the field. In the center of the formation, a standard bearer plants his feet and steadies a tall pole crowned with a bronze eagle. The metal catches the rising sun, its wings spread wide as if poised for flight. Leather streamers hang below, marked with the legion's number and honors.

The men do not look first to their general. Their eyes fix on the eagle. It marks their place in the line and binds them to Rome. When the standard shifts forward, the ranks move with it. When it pauses, the line holds firm. The bearer's hands tighten around the shaft as distant horns sound. In this moment, authority stands not only in shouted command, but in the gleam of metal lifted above the field.

An Army Reforged for Distant Wars

In the late second century BCE, Rome faced threats on multiple fronts. Wars in North Africa and northern Europe strained the old citizen militia system. Gaius Marius, elected consul in 107 BCE, reorganized the army to meet these pressures. He opened enlistment to landless men and shaped legions into standing forces loyal to their commanders and to Rome. Equipment became more standardized. Training grew stricter and more uniform.

Each legion needed clear structure and identity as it marched far from Italy. Units were divided into cohorts and centuries, each with its own officers. Standards marked these divisions in battle and camp. The legionary eagle, adopted as the single emblem for each legion, stood at the center of this new order. Carried by a chosen soldier, the standard gave visible form to command within an army that now fought year round across distant provinces.

Forged in Bronze and Oath

The legionary eagle, or aquila, was crafted from bronze or silver and fixed to the top of a long wooden shaft. Skilled metalworkers shaped the bird with spread wings and sharp beak, often adding fine details to feathers and talons. The shaft was fitted with metal grips and crossbars to steady it in wind and movement. Leather straps and small decorations marked the legion's identity. Each eagle was unique to its unit and guarded with care.

The standard bearer, known as the aquilifer, was chosen for strength and reliability. He received higher pay and carried the eagle into battle and on parade. The creation of a new legion included the forging or presentation of its eagle. Once assigned, the object became more than equipment. It was stored in a shrine within the legion's camp, under guard, when not in use. Only authorized hands could touch it.

Eyes Fixed on the Standard

When a legion assembled for the first time under its new eagle, soldiers gathered around the standard in silence. Veterans watched closely as the aquilifer lifted the pole upright. The bronze bird rose above helmets and shields, catching light and drawing steady attention. For recruits, the eagle marked entry into a brotherhood that carried Rome's authority beyond Italy. The object was visible proof of belonging and duty.

Commanders also understood its weight. Before battle, they addressed troops with the eagle in view, knowing that loyalty gathered around it. Yet uncertainty lingered. A standard could fall in chaos. Enemies targeted it, aware that its loss could shake morale. In moments of confusion, soldiers looked to the eagle for direction. If it wavered, hesitation spread.

Authority in the field rested not only in orders shouted across ranks, but in the steady presence of the raised standard.

Fear of Loss in Foreign Lands

As Roman legions pushed deeper into Gaul, Germania, and the Near East, the risk of losing a standard grew sharper. Reports of ambush and sudden attack reached Rome. Senators and generals knew that the capture of an eagle could damage more than a single unit. It could weaken the image of Roman control. Rival leaders watched closely for signs of weakness. The fate of each standard carried weight far beyond the battlefield.

Within the ranks, pressure settled on the aquilifer. He stood near the front in combat, visible to both friend and foe. Fellow soldiers expected him to protect the eagle at all cost. Religious meaning also deepened the concern. The eagle was linked to Jupiter, Rome's chief god. Priests performed rituals before campaigns, invoking divine favor over the standards. Loss would not only shame a legion. It would suggest that the gods had withdrawn support.

Guarded Emblems Within the Camp

Roman camps were laid out with strict order, and the standards were placed in a central shrine known as the sacellum. Archaeological remains from permanent forts show foundations for these sacred spaces within headquarters buildings. When not carried into the field, the eagles stood inside these shrines under guard. Access was limited to officers and appointed soldiers. Ritual offerings were sometimes made before them, linking military discipline with religious practice.

Ancient historians record efforts to recover lost standards with great urgency. After the defeat in the Teutoburg Forest in 9 CE, several eagles were captured by Germanic forces. Roman sources describe campaigns launched in later years to retrieve them. While these accounts highlight honor and revenge, surviving inscriptions confirm that recovered standards were displayed publicly in Rome. The state treated the objects as symbols requiring strict protection and official handling.

Honor Carried on One Man's Shoulders

The aquilifer bore more than the weight of bronze and wood. In battle, he stood where fighting was thickest, expected to hold the eagle high even under heavy attack. If the line faltered, he could not retreat easily without drawing notice. Ancient accounts describe standard bearers who fell while defending their eagles, their deaths remembered in inscriptions and later histories. Survival brought honor, but it also carried strain.

For commanders, the loss of a standard could stain a career. After disasters such as the Teutoburg Forest, reputations suffered in Rome. Families of fallen soldiers endured both grief and public shame tied to captured eagles. Veterans spoke quietly of those campaigns, aware that the missing standards marked unfinished business. Even in peacetime, the memory of loss lingered in legionary circles, shaping how soldiers measured courage and loyalty.

Standards Cataloged and Reclaimed

Over time, the Roman state formalized procedures for guarding and honoring the standards. Permanent legions stationed along frontiers kept detailed rosters and inventories that included their eagles. The standards were stored within headquarters buildings under assigned watch. During transfers or reassignments of units, records tracked movement of equipment and symbols alike. The eagle did not change hands casually. Its custody followed rank and formal appointment.

When standards were lost in battle, emperors treated recovery as a state matter. Diplomatic negotiations and military campaigns sought their return. After Augustus secured the return of standards captured by the Parthians in 20 BCE, inscriptions and monuments recorded the event. The recovered eagles were placed in temples in Rome under official supervision. Their handling became part of imperial ceremony, managed through established administrative channels rather than private initiative.

Echoes of Lost Eagles

Years after the disaster in the Teutoburg Forest, Roman forces continued to search for the missing standards of the defeated legions. Campaigns under Germanicus in the early first century CE pushed into the forests where the earlier army had fallen. Ancient sources describe the recovery of at least one eagle during these expeditions. The effort showed that memory of loss endured within military and political circles long after the battle itself.

Yet not every standard was quickly returned. Some remained in enemy hands for years, perhaps displayed as trophies. In Rome, the absence carried weight. Veterans spoke of the fallen legions in subdued tones. New recruits trained under different eagles, aware of past failures. The bronze birds lifted above fresh formations, but the memory of those once captured lingered within the ranks and the Senate alike.

Fragments in Earth and Stone

Few legionary eagles survive intact today, yet traces of Roman standards appear in reliefs, coins, and carved monuments across former imperial lands. The Arch of Titus in Rome shows soldiers carrying standards with careful detail. Military diplomas and inscriptions record the names of aquilifers and the legions they served. Archaeological digs at frontier forts have uncovered parts of standard fittings, metal attachments, and decorative elements that once crowned poles raised in battle.

No complete eagle from the early Empire has been found in its original setting, yet the physical evidence aligns with written accounts. The weight of bronze fragments, the wear on metal grips, and the placement of shrine foundations within camps reveal how closely guarded these objects were. Today, museum cases hold what remains of fittings and reliefs. The bronze bird that once guided a legion survives in carved stone and scattered metal, its authority preserved in fragment and image.

Chapter 5
The Pillar That Carried the King's Voice — Ashoka's Edicts and Moral Rule

In the third century BCE, under a wide Indian sky, laborers heave against thick ropes to raise a towering sandstone pillar from the dust. The shaft is smooth and pale, cut from a single block quarried far from this village road. Sweat runs down bare arms as wooden scaffolds groan. Slowly, the pillar tilts upright. At its top rests a carved capital of lions, their mouths open, their bodies facing the four directions.

Imperial officers step forward once the base is set firm in the earth. Stonecutters unpack iron tools and begin to mark lines across the polished surface. Each strike of the chisel leaves deep, even characters in the stone. The script runs in steady rows meant to be read by officials and travelers who pass along the road. Farmers pause at a distance, watching the pillar rise above fields and trees. The king's words are about to stand in open light.

An Empire Bound by Roads and Conquest

Before Ashoka ordered pillars set across his realm, the Mauryan Empire had already stretched from the Indus Valley to the eastern plains. Chandragupta Maurya built the empire through war and alliance in the late fourth century BCE. His son Bindusara extended its reach southward. By the time Ashoka came to the throne around 268 BCE, he inherited a vast territory tied together by roads, garrisons, and provincial governors.

The empire was ruled from Pataliputra, a capital filled with officials who kept records and oversaw taxes, trade, and law. Royal orders moved along river routes and highways. Governors known as mahamatras carried out the king's commands in distant districts. After the bloody conquest of Kalinga, Ashoka faced the challenge of ruling a population that had felt the force of Mauryan arms. Authority depended not only on soldiers, but on visible signs of royal presence across the land.

Sandstone Raised and Words Cut Deep

The pillars were carved from fine grained sandstone quarried at sites such as Chunar near the Ganges. Workers cut massive blocks from the earth and shaped them into long, cylindrical shafts. The surfaces were polished to a high sheen, giving the stone a smooth finish that reflected light. Transporting the pillars required careful planning. They were hauled overland and along rivers, moved by teams of laborers using sledges, rollers, and boats.

Once positioned upright at key roads, cities, or sacred sites, masons began to carve the edicts. The text was written in Prakrit and incised in the Brahmi script, with some versions in other regional scripts. Skilled stonecutters followed prepared drafts, spacing the lines evenly across the polished surface. The letters were cut deeply to withstand weather. Each pillar carried a version of the king's message, fixed in stone under official supervision.

Villagers Beneath the Lions

When the first pillars rose beside busy roads and near market towns, travelers stopped to look up at the polished shafts and carved lions. Local officials gathered to hear the inscriptions read aloud. Many villagers could not read the script themselves, so scribes or officers recited the king's words in public. The edicts spoke of moral conduct, restraint, and care for subjects. Some listeners nodded in approval, hearing a call for order and fairness.

Others listened with quiet caution. The same ruler who now urged compassion had once led armies into Kalinga with heavy loss of life. Farmers and merchants had seen imperial tax collectors and soldiers in their districts. The pillars stood firm and gleaming, but questions lingered about how these moral commands would shape daily rule. Authority felt both present and distant as the carved words echoed across open fields.

Governors Under a Moral Gaze

As more pillars were set across the empire, provincial governors felt the weight of the king's public words. The edicts called for fair treatment of prisoners, restraint in punishment, and respect for different religious groups. These instructions were carved in stone where merchants, monks, and farmers could hear them read aloud. Officials who had long ruled through firm discipline now faced a standard displayed in open daylight.

Religious leaders also watched closely. Ashoka's support for Buddhist communities was known, yet the edicts addressed Brahmins, ascetics, and other groups as well. Temples and monasteries depended on royal favor and local order. If governors failed to uphold the moral tone of the inscriptions, complaints could travel back to Pataliputra. In administrative halls, conversations grew guarded. The pillars stood silent, but their presence sharpened attention on how power was exercised in distant provinces.

Reading the Edicts in Open Ground

Imperial officers were tasked with ensuring that the edicts were understood. Some inscriptions mention officials appointed to promote moral conduct and to report back to the king. These mahamatras traveled through districts, explaining the king's instructions and hearing local concerns. The pillars themselves remained fixed in place, but the meaning of the carved lines required spoken interpretation for many listeners.

There is no ancient record that the texts were altered once cut into stone. Later legends sometimes described Ashoka as issuing sweeping spiritual reforms beyond what the inscriptions state. The surviving edicts, however, present measured language about duty, respect, and welfare. Access to the pillars was not restricted by walls or guards in the way temple objects were, yet their wording was controlled at the time of carving. Once set, the text stood beyond revision, its surface closed to further change.

Officials Measured by Public Words

Provincial officers who failed to follow the tone of the edicts risked more than quiet criticism. Complaints could reach the capital through traveling inspectors. An official known for harsh punishment or corruption could face removal from post. The pillars made the king's expectations visible in every district. Administrators now worked under a standard carved in stone, aware that subjects could point to the words if treated unjustly.

Religious leaders also felt strain. Ashoka's calls for respect among sects required restraint in public rivalry. Monks and Brahmins who had competed for influence had to weigh their speech more carefully. Those who appeared too eager to claim royal favor risked resentment from other groups. In towns where the pillars stood, conversations shifted. The polished stone reflected more than sunlight. It reflected reputations shaped by how closely one followed the king's declared path.

Orders Sent Along the Royal Roads

Over the years, additional edicts were carved in different regions of the empire. Some pillars carried updated messages that addressed new concerns, such as treatment of animals or the conduct of officials. Couriers moved between Pataliputra and provincial centers with written instructions. Local administrators recorded compliance in reports sent back to the capital. The process tied distant districts to a steady stream of royal communication.

The pillars themselves were not relocated once erected. They remained where they were set, marking roads, borders, and sacred spaces. Maintenance fell to local authorities, who cleared ground around the bases and ensured the inscriptions remained visible. No record shows that later Mauryan rulers removed the pillars, though new orders were issued through other means. The carved stone stood as a fixed channel of command within an evolving administrative system.

Pillars Standing After the King

After Ashoka's death in the early second century BCE, the Mauryan Empire began to weaken. Regional leaders asserted greater control, and the central authority from Pataliputra faded. Yet the pillars remained where they had been set. Travelers continued to pass beneath the carved lions. The inscriptions did not change with shifting power. They stood as reminders of a ruler who had spoken across his realm in stone.

In some areas, later builders reused pillars or moved them short distances, though many stayed in place for centuries. The political structure that raised them did not endure, but the polished shafts outlasted the officials who once read them aloud. Generations saw the script without fully understanding it. The authority that first commanded attention through these edicts had receded, yet the silent columns continued to mark the landscape with a king's voice carved into stone.

Letters Brought Back to Light

Centuries after the Mauryan Empire faded, many pillars still stood across northern India, though their script had long been unread. In the nineteenth century, scholars such as James Prinsep studied the inscriptions carefully, comparing repeated signs across different pillars and rock edicts. By tracing patterns and matching names known from other sources, they gradually deciphered the Brahmi script. The carved lines that had been silent for generations began to speak again.

Archaeologists documented the polish of the sandstone, the precision of the letters, and the transport routes from the Chunar quarries. Some pillars had been broken or moved, yet their surfaces still bore the original cuts of Mauryan masons. Today, visitors stand beneath the lion capitals preserved in museums or at original sites. The stone still gleams in sunlight, its carved words carrying across time from the reign of Ashoka.

Chapter 6
The Jade That Spoke for Heaven — The Heirloom Seal of the Realm

In 221 BCE, inside the capital of Xianyang, oil lamps cast steady light across lacquered pillars and dark wooden beams. Courtiers stand in ordered rows as a block of pale jade rests on a low table draped in silk. The stone has been carved into a square seal, its surface polished smooth. Along its base, small characters have been cut with care. The room is quiet except for the faint scrape of fabric and the soft breath of waiting officials.

An attendant brings forward a document prepared on bamboo slips. A senior court officer lifts the jade seal with both hands. He presses it into thick red paste, then lowers it firmly onto the surface of the text. When the seal rises, a crisp imprint remains. Outside the palace, former rival states adjust to new rule. Inside, the jade has marked the will of the First Emperor with a sign meant to carry the mandate of Heaven.

A Realm Forged from Warring States

Before the jade seal marked imperial orders, China had endured centuries of conflict among rival kingdoms. During the Warring States period, armies marched across plains and river valleys, and alliances shifted without warning. The state of Qin, hardened by reform and strict law, rose in strength. Under King Zheng, Qin armies defeated Han, Zhao, Wei, Chu, Yan, and Qi in steady succession.

In 221 BCE, Zheng declared himself Qin Shi Huang, the First Emperor. He ordered the standardization of weights, measures, currency, and even the written script. Roads were built to link distant regions. Officials replaced former nobles in conquered lands. Authority flowed from the capital at Xianyang through appointed administrators. In this newly unified realm, documents required clear confirmation that they carried the emperor's command. The seal would serve as that visible mark of rule.

Jade Cut for an Emperor

According to early historical records, the seal was carved from a famous piece of jade known as the Heshibi. The stone was said to have passed through noble hands before reaching the Qin court. Once unified rule was declared, imperial craftsmen shaped the jade into a square block with a carved handle. The surface was polished until it gleamed pale under torchlight. Jade was prized not only for beauty, but for strength and rarity.

An inscription was cut into the base in the small seal script standardized under Qin rule. The characters declared that the emperor had received the mandate from Heaven and would enjoy long prosperity. Skilled artisans incised the lines in reverse so that each impression would appear correctly on silk or bamboo. Once finished, the seal was kept within the palace, handled only by trusted officials authorized to mark imperial documents.

A Mark That Carried Heaven's Claim

When the first imperial decrees bore the imprint of the jade seal, court officials examined the red mark with careful eyes. The characters were clear and firm, linking the emperor's order to a claim of Heaven's mandate. Ministers bowed as documents were sealed and dispatched to distant commanderies. Messengers carried the bamboo slips along newly built roads, the imprint serving as proof that the words came from Xianyang.

In conquered regions, former nobles and local leaders saw the mark on orders that replaced their own authority. Some accepted the new seal as the final sign of Qin dominance. Others read the claim of Heaven with guarded thoughts. The unification had been swift and harsh. Armies had burned cities and disarmed rivals. The seal's impression fixed the emperor's will in red paste, yet beneath formal compliance lingered quiet uncertainty about how long such power would endure.

Courtiers Under the Emperor's Shadow

As Qin rule tightened, ministers at court felt the weight of decisions stamped with the jade seal. Legal reforms were strict, and punishment for disobedience could be severe. When a document carried the red imprint, it left little room for hesitation. Officials who questioned orders risked suspicion. The seal's presence on edicts reinforced the reach of the central government into every province.

Scholars and former nobles from defeated states also sensed pressure. Some had lost lands and titles when Qin administrators replaced hereditary rulers. The emperor's claim to Heaven's mandate, pressed into each decree, challenged older traditions tied to regional loyalty. Conversations in private halls grew cautious. Religious rites and court ceremonies acknowledged the emperor's authority, yet memories of earlier dynasties remained alive. The jade seal stood at the center of this shifting balance, a symbol watched by friend and rival alike.

Guarded in the Inner Palace

The jade seal was kept within the imperial palace and handled under strict protocol. Court records from later dynasties describe procedures for presenting the seal during formal audiences. It was not left in open view. Trusted attendants stored it in guarded chambers and brought it forward only when the emperor or a high minister required it for official use. The act of sealing a document followed ritual order, reinforcing the authority behind each command.

Later histories, written under the Han and subsequent dynasties, recount stories about the seal's journey through rebellion and court intrigue. These accounts describe it passing from ruler to ruler as regimes rose and fell. While some details cannot be verified by surviving objects, early chronicles consistently treat the seal as a central emblem of sovereignty. Control of the physical jade block meant control over the right to mark imperial decrees.

Lives Bound to a Single Imprint

Court officials who held temporary custody of the seal carried heavy responsibility. A misplaced seal or unauthorized imprint could bring severe punishment under Qin law. Ministers who presented sealed orders to regional governors stood exposed if those commands led to unrest or hardship. Their names were tied to documents that reshaped taxes, labor drafts, and military service. The red mark did not shield them from blame.

During the collapse of the Qin dynasty, possession of the seal became a matter of survival. Generals and rival claimants to the throne sought objects that proved legitimate rule. Those entrusted with safeguarding the seal faced danger if factions changed power within the capital. Loyalty could shift overnight. In such moments, the jade block was more than a tool of state. It was a prize that placed its guardian at the center of political struggle.

Passed Between Dynasties

After the fall of the Qin dynasty in 206 BCE, control of the seal shifted amid civil war. Early Han records describe how the jade block came into the possession of Liu Bang, who founded the Han dynasty as Emperor Gaozu. The seal was incorporated into new court rituals, used to mark imperial edicts in the name of the Han ruler. Its presence linked the new regime to the claim of Heaven's mandate first asserted under Qin.

Over time, official chronicles recorded each transfer of the seal during periods of upheaval. When emperors were overthrown or rival courts formed, the seal was treated as a key object to secure. Protocols governed its storage and presentation in palace halls. Court historians noted when it was lost, recovered, or carried into exile. Administrative order depended not only on armies and officials, but on control of the physical seal within the capital.

A Symbol Sought in Times of Fracture

Centuries after the Qin unification, the Heirloom Seal remained tied to claims of legitimate rule. During periods of division, rival leaders sought to obtain it to strengthen their authority. Historical chronicles from the Three Kingdoms era describe negotiations and military moves shaped in part by the seal's location. Possession of the jade block was treated as proof that Heaven favored one ruler over another.

Yet the seal's path grew uncertain over time. By the late Tang and early Song periods, records became less clear. Some accounts state that it was lost during upheaval in the tenth century. Later emperors commissioned new seals to mark their edicts, but the original jade was no longer securely documented. The absence left space for doubt. A dynasty could proclaim Heaven's mandate, yet the famed seal that once confirmed it no longer stood in court.

Records Without the Jade

No confirmed physical trace of the original Heirloom Seal survives today. What remains are descriptions preserved in early chronicles such as the Shiji and later dynastic histories. These texts record the seal's inscription, its material, and the ceremonies in which it was used. Scholars compare these written accounts with surviving imperial seals from later dynasties to understand how such objects were crafted and guarded.

Archaeologists have uncovered jade seals from tombs and administrative sites, showing the size, weight, and carving style common in early imperial China. None can be securely identified as the lost heirloom. Museums display these later seals behind glass, their bases still cut in mirror script. The famed jade block that once pressed red authority onto bamboo slips has vanished. What endures is the record of how one object marked the right to rule under Heaven.

Chapter 7
The Stone That Guarded the Fields — Kudurru Boundary Markers and Land Authority

Under a wide Mesopotamian sky in the twelfth century BCE, farmers stand at the edge of a field newly measured and marked. The soil is dry, split by heat and wind. Laborers guide a dark stone into a narrow pit dug along the boundary line. The block is heavy, carved on one face and left rough on the back. Along its upper edge, small symbols of gods, crescent, star, and horned crown, have been cut into the surface.

A royal scribe kneels beside the stone and reads the inscription in a steady voice. The text names the Kassite king who granted the land and lists the official witnesses present. It describes the field's size, the canals that border it, and the curses laid upon anyone who shifts the marker. As the stone settles into the earth, farmers watch in silence. Authority has taken shape in carved rock pressed into their soil.

Fields Under Royal Grant

During the Kassite period, which began in the sixteenth century BCE, Babylonia was ruled by kings who traced their authority through both conquest and alliance. The Kassite dynasty maintained control over cities, temples, and farmlands along the Tigris and Euphrates rivers. Kings rewarded loyal officials, military officers, and temple servants with grants of land. These grants often included fields, orchards, and access to water channels that made farming possible in the dry plain.

Royal authority over land did not rest on memory alone. Temple estates, private holdings, and crown lands were carefully recorded on clay tablets. Disputes over boundaries could disrupt tax collection and labor duties. In this setting, a permanent marker carried weight. When a king granted land, the decision had to be fixed in a form that could withstand challenge from neighbors, heirs, or rival claimants. The boundary stone stood as that fixed point in the open field.

Carved Deeds in Stone

Kudurru stones were cut from durable limestone or dark diorite and shaped into upright slabs. One face was smoothed for carving, while the back often remained rough. At the top, artisans carved symbols linked to major gods, such as Shamash, Enlil, or Ishtar. These emblems served as divine witnesses to the grant. Below the symbols, trained scribes drafted the legal text in Akkadian cuneiform before stonecutters incised it into the surface.

The inscription named the king, the beneficiary of the land, and the precise measurements of the field. It described canals, neighboring plots, and tax exemptions tied to the grant. The text also listed curses calling down divine punishment on anyone who altered the boundary. Once completed, the stone was set either in the field itself or placed in a temple for protection, depending on local practice and royal instruction.

Neighbors at the Edge of the Field

When a kudurru was raised, nearby landholders came to witness the act. The scribe read the grant aloud so that boundaries were clear in public hearing. Farmers listened for references to canals, paths, and shared grazing areas. Some nodded as measurements matched what they knew. Others shifted uneasily if the new lines seemed to press against older claims passed down through families.

Temple officials also took notice. If the grant included tax exemption or relief from labor duty, it affected temple revenue and obligations. The carved curses invoked powerful gods, which added weight to the marker's authority. Yet carved words did not silence every doubt. In private, neighbors measured distances again and traced the edges of fields with their eyes. The stone stood firm, but daily life around it continued to test its claim.

Boundaries That Tightened the Plain

As more grants were issued, tension grew among those whose land bordered royal gifts. A single stone could shift access to water or reduce shared grazing ground. Village elders worried that too many exemptions might weaken communal labor systems. Canal maintenance required joint effort. If one holder claimed relief from duty under royal favor, others carried heavier loads.

Temple authorities felt pressure as well. Some grants reduced taxes that once flowed into sacred estates. Priests could not ignore a king's decision carved in stone, yet they had to protect temple resources. Complaints traveled quietly through market stalls and along irrigation ditches. The symbols of the gods carved at the top of each kudurru stood as warning, but they also drew attention. Every boundary stone marked not only land, but a balance of obligation under watchful eyes.

Reading the Curses, Guarding the Claim

When disputes arose, officials referred back to the carved text on the kudurru. Scribes compared the stone inscription with clay tablet copies kept in temple archives. Surviving examples show that many kudurru stones were stored within temples rather than left exposed in fields. This practice protected them from damage and allowed priests to oversee their use as legal evidence. The stone's text could be read aloud in hearings to confirm the king's grant.

Later tradition sometimes treated the curses on the stones as purely symbolic, yet the inscriptions were formal legal documents. The threats invoked named gods and detailed specific punishments. Access to the physical stone was controlled by temple staff. Not every villager could approach and inspect it freely. The authority of the boundary rested not only in its placement in soil, but in its guarded preservation within sacred space.

Lives Measured by the Marker

For the beneficiary of a royal land grant, the kudurru brought both privilege and risk. A man who received tax exemption or expanded fields gained status in his village. Neighbors might seek his favor or resent his rise. If crops failed or canals shifted, blame could fall on the one whose boundary had been altered by royal order. The stone in the field marked him as favored, and that mark drew attention.

For those who lost land or shared access, the cost was practical and personal. Families who had farmed a strip for years could find it reduced. Appeals to local elders carried less weight once a king's name stood carved in stone. Tension could isolate households on either side of the marker. The gods invoked in the curses watched over the boundary, yet daily strain settled among the people who worked the soil around it.

Grants Recorded and Guarded

Over time, the Kassite administration developed routine steps for confirming land grants. Once a kudurru was carved, a corresponding record was often entered on clay tablets and stored in temple or palace archives. Officials could consult these records if boundaries were questioned. The stone and the tablet worked together, one set in durable rock, the other preserved among written accounts.

Some kudurru stones were placed inside temple precincts rather than in open fields. This practice reduced the risk of damage or tampering. Access was shaped by temple rules, and officials oversaw formal readings when disputes arose. If land passed to heirs, the original grant remained tied to the stone's inscription. No record shows casual removal once a boundary was confirmed. The process favored stability, with royal authority anchored in both soil and sanctuary.

Stones That Outlasted Their Kings

Decades after a kudurru was set in place, the original king who issued the grant was often gone. New rulers sat on the throne in Babylon. Yet the boundary stone still named the earlier monarch and repeated his decision. Heirs relied on the carved text to defend their claim when neighbors or officials questioned it. In some cases, later rulers confirmed older grants, leaving the original stone undisturbed.

Political change, however, did not always protect the landholder. When foreign powers entered Babylonia or when dynasties shifted, older grants could be challenged. Some stones were moved into temple storage for safety. Others were carried off as trophies by invading armies. The carved curses remained visible, but they could not prevent every disruption. The boundary stood in stone, yet the balance of power around it continued to shift with time.

Stones in Museum Light

Many kudurru stones survived because they were stored in temples or later buried in palace ruins. In the nineteenth and early twentieth centuries, archaeologists excavating sites such as Susa uncovered boundary markers carried off in ancient conflicts. The stones were found broken in some cases, yet their inscriptions and divine symbols remained clear. Scholars translated the Akkadian text and compared it with surviving clay tablets from Kassite Babylonia.

Today, these boundary stones stand in museum galleries far from the fields they once marked. The carved emblems of gods remain sharp against dark stone. Visitors read translations that list land measurements and curses laid upon violators. The soil that once pressed against their bases is gone, replaced by polished floors and glass cases. Yet the stones still bear the weight of royal authority cut into their surface, fixed in rock and carried across centuries.

Chapter 8
The Ring That Sealed the Crown — Egyptian Signet Rings and Royal Authority

In the cool shade of a palace courtyard in New Kingdom Egypt, a goldsmith leans over a low table scattered with tools and scraps of metal. Thin sheets of gold catch the light as he shapes a small ring with careful hands. Beside him rests a carved scarab seal, its underside etched with tight hieroglyphs that spell the throne name of the reigning pharaoh. The air smells faintly of heated metal and damp clay.

A senior official steps forward holding a rolled papyrus tied with linen cord. A small lump of wet clay is pressed flat over the knot. The official slides the ring from his finger and lowers the scarab face into the clay. He lifts it again, leaving a crisp imprint. The mark is small, barely wider than a thumb, yet it carries the king's name. Around them, scribes and attendants watch in silence as authority is fixed with a single press of gold.

Orders Moving Along the Nile

In the Middle and New Kingdom periods, Egypt was governed through a structured chain of command. The pharaoh stood at the top as divine ruler, but daily administration passed through the vizier and a network of scribes and regional governors. Tax records, labor assignments, and temple accounts were written on papyrus and stored in archives. Officials traveled between Memphis, Thebes, and provincial centers carrying documents that directed work along the Nile.

Each order required proof that it came from authorized hands. Clay seals secured doors, storage jars, and bundles of papyrus. Signet rings bearing royal names or official titles provided that proof. In a system where written records guided irrigation, grain storage, and temple offerings, the seal confirmed authenticity. Without the correct imprint, a document could be questioned or refused. Authority depended not only on words written in ink, but on the mark pressed into clay.

Gold, Stone, and Carved Names

Signet rings in ancient Egypt were crafted from gold, electrum, or other precious metals shaped into bands that fit securely on a finger. At the center, artisans set a carved scarab or flat bezel made from stone such as steatite or carnelian. The underside of the scarab was engraved in reverse with hieroglyphs spelling a royal throne name or the title of a high official. This careful carving ensured that the impression appeared correctly when pressed into clay.

Goldsmiths and stone carvers worked in palace or temple workshops under supervision. The engraving required steady hands and knowledge of formal script. Once completed, the ring was assigned to its bearer and used to seal papyrus documents, storage jars, and doors. The small carved surface became the point where royal identity met written command, fixed in a mark no larger than a thumbprint.

A Mark Recognized in Market and Temple

When sealed documents arrived in provincial offices, scribes examined the clay impressions before reading the text. The carved name of the pharaoh or a trusted official signaled that the order carried weight. Storehouse keepers broke jar seals only after checking the imprint. In temple courts, priests accepted deliveries of grain or oil once the mark matched known authority. The small scarab image became familiar across towns along the Nile.

Yet hesitation could arise if a seal seemed unclear or unfamiliar. A poorly pressed imprint or a name rarely seen in a district might raise quiet questions.

Officials sometimes compared impressions with known seals kept in archives. In times of transition between rulers, the appearance of a new throne name on rings required careful attention. The imprint confirmed command, but it also demanded recognition from those who received it.

Trust Bound to a Finger

As Egypt's administration grew more complex, the number of officials granted personal seals increased. Viziers, treasurers, and regional governors wore rings that carried their titles. With each new appointment, the circle of those able to authorize orders widened. Senior administrators worried about misuse. A forged impression or an unauthorized document could redirect grain shipments or labor crews. The pressure to guard one's ring grew intense.

Temple authorities also felt strain. Seals secured storerooms filled with offerings and tax goods. If a seal was broken without proper record, suspicion followed. Priests and scribes watched closely when new officials arrived bearing unfamiliar signet rings. In periods of succession, when a new pharaoh's name replaced the old, confusion could spread through the bureaucracy. Authority rested in small carved surfaces that traveled on human hands, exposed to risk each day.

Comparing Impressions in Clay

Egyptian scribes developed routine methods to verify seals. In administrative offices, broken clay sealings were often stored and compared with known impressions. Archaeologists have uncovered hundreds of such sealings at sites like Amarna and Thebes, many still bearing the clear imprint of royal or official names. These fragments show that impressions were examined closely before goods were released or documents accepted.

There is no ancient record describing a formal handbook for seal verification, yet surviving archives suggest careful oversight. When a pharaoh died, new throne names were carved onto fresh rings, and older seals were set aside. Some rings were buried with their owners in tombs, removed from circulation. Later stories sometimes speak of magical power tied to scarabs, but the surviving sealings show practical use. The imprint served as controlled proof within daily administration.

A Ring That Could Ruin a Career

For an official entrusted with a signet ring, loss or misuse carried serious danger. If a sealed document proved false or unauthorized, suspicion fell first on the ring's bearer. Records from later periods show that officials could be removed from office for corruption or failure in duty. A broken seal discovered without record could lead to inquiry and public shame within the court.

Servants and scribes who handled sealed goods also faced strain. If a jar arrived with a damaged impression, they had to decide whether to accept it or report the fault. A wrong choice could cost them position or favor. The ring was small enough to hide in a sleeve or lose in crowded halls. Each time it left a mark in clay, it bound a person's name and status to the outcome of that order.

Seals Cataloged and Replaced

Over time, the Egyptian administration developed clear routines for issuing and retiring signet rings. When a new official was appointed, a ring bearing his title was prepared under supervision. Records in temple and palace archives noted the transfer of authority. Upon death or removal from office, the ring was often reclaimed or placed in a tomb, limiting further use. This process reduced confusion during transitions of power.

Clay sealings were sometimes stored in storerooms or archive rooms after documents were opened. Excavations at sites such as Amarna reveal large deposits of broken seal impressions that once secured doors and containers. These fragments acted as records of prior transactions. If disputes arose, officials could refer to stored sealings as proof that goods had passed through authorized hands. The system relied on careful custody of both ring and impression.

Names That Outlived Their Bearers

Years after a pharaoh's reign ended, his throne name continued to appear on scarabs buried in tombs or left in abandoned archives. Officials who once carried rings with that name were gone, yet their sealings remained pressed into hardened clay. In some cases, later rulers reused older names or honored earlier kings, keeping familiar symbols in circulation.

During periods of political change, confusion could arise when new seals replaced old ones. A document bearing a former ruler's name might prompt hesitation if discovered after succession. Some sealings were broken and discarded as administrations shifted. Yet others remained in sealed rooms until uncovered centuries later. The small carved surfaces that once confirmed royal command survived in fragments, scattered among ruins where authority had long since passed.

Impressions Preserved in Dust

Archaeologists working in Egypt have uncovered thousands of clay sealings in palace ruins, temple storerooms, and administrative quarters. Many still bear clear impressions of scarab seals with royal names and official titles. At sites like Amarna, broken lumps of clay were found in heaps near archive rooms, each once pressed against papyrus ties or jar lids. Though the papyrus has decayed, the imprint remains sharp.

Museums now display gold signet rings and carved scarabs recovered from tombs and settlements. The hieroglyphs are still legible, cut in reverse so their impressions would appear correctly in clay. Scholars match these seals with known reigns and offices, tracing lines of authority across centuries. The ring that once pressed into soft clay survives in metal and stone, its small carved face still carrying the name of a king.

Chapter 9 The Rock That Marked the Border — The Behistun Inscription and Imperial Claim

High above an ancient road in western Iran, workers cling to narrow ledges carved into a pale cliff face. The rock rises steep and sheer over the route that links Mesopotamia to the Iranian plateau. Wind presses against loose garments as chisels strike stone. Each blow sends chips falling toward the road below. Slowly, a royal relief emerges from the rock. A crowned king stands firm, one foot placed upon a fallen rival, while bound figures line up behind him.

Travelers on the road pause and shade their eyes to look upward. The inscription is set far beyond easy reach. Lines of text spread across the cliff in ordered columns, carved in more than one language. Dust drifts through the air as masons steady themselves on wooden supports. The scene rises above the caravan path, fixed into living rock, declaring that this land now stands under the rule of Darius.

A Throne Claimed in Turbulent Days

When Cambyses II died in 522 BCE, the Persian Empire faced sudden unrest. A man claiming to be Bardiya, brother of Cambyses, seized control. Many provinces accepted his rule, uncertain of the truth. Soon after, Darius, a noble of the Achaemenid line, led a group of conspirators who killed the claimant. Darius declared that the man was an impostor named Gaumata. The empire was not yet stable.

Rebellions broke out across distant regions, from Babylon to Egypt. Local leaders declared independence or named rival kings. Darius moved quickly with loyal forces, fighting campaign after campaign to restore control. Messages had to travel along major roads linking the heartland of Persia to Mesopotamia. In this atmosphere of challenge and doubt, authority required visible proof. A ruler's claim needed to stand where merchants, soldiers, and officials could not ignore it.

Chisels on a Sheer Cliff

Darius chose a limestone cliff at Behistun, near a well traveled royal road, as the site for his inscription. The face of the rock was cut smooth in a large panel high above the ground. Workers built scaffolding against the cliff, carving footholds into the stone to steady themselves. The relief of the king and his defeated rivals was shaped first, with careful attention to clothing, weapons, and posture.

Below and beside the relief, scribes prepared the text in Old Persian, Elamite, and Babylonian cuneiform. Drafts were likely written on clay or parchment before being transferred to the rock. Stonecutters then incised the characters into the smoothed surface. The work required coordination between scribes and masons to keep lines straight and legible. Once carved, the inscription became part of the cliff itself, beyond easy reach of alteration.

Eyes Raised to the Cliff

Caravans moving along the royal road could not miss the towering relief above them. Merchants and messengers stopped to study the carved figures, pointing at the king standing over bound rebels. Those who could read cuneiform examined the lower panels more closely. The presence of three languages signaled that the message was meant for different peoples within the empire. The scale alone commanded attention.

Local officials traveling between provinces recognized the names of revolts listed in the text. Some had lived through the unrest described on the rock. Hearing Darius's version carved into stone may have brought relief to loyal supporters. Others, whose regions had once backed rival leaders, might have read the account with guarded thoughts. The inscription stood firm above the road, yet memories of recent conflict lingered among those who passed beneath it.

Names Carved as Warning

As word spread of the Behistun inscription, provincial governors understood its message. The text listed rebels by name and described their defeat in detail. Officials who had once hesitated in their loyalty saw those names fixed in stone beneath the king's image. The relief showed bound figures led before Darius, each labeled with the region he claimed. The display left little room for doubt about the fate of challengers.

Religious leaders and local elites also felt the weight of the carving. The inscription declared that Darius ruled by the favor of Ahura Mazda. Temples that depended on royal support watched closely as the king's authority was presented as divinely backed. For communities that had briefly followed other claimants, the public record served as a reminder. The cliff face rose above the road, silent yet constant, pressing its account upon all who traveled beneath it.

Reading from the Road Below

The Behistun inscription was carved high above the ground, making close study difficult. Travelers could see the relief clearly, but reading the text required knowledge of cuneiform and a position close enough to trace the lines. There is no record that ladders or platforms were left in place for public use once the carving was finished. The height itself limited access and helped protect the surface from damage.

Copies of the text were later made on clay tablets and circulated within the empire. Surviving fragments found in Babylon confirm that scribes recorded the Old Persian account in more portable form. These copies allowed officials to consult the king's version of events without climbing the cliff. Later traditions sometimes described the inscription as meant only for the gods, but its placement above a busy road suggests it was also intended for human witness.

Lives Named in Stone

For those listed as rebels on the cliff face, their names were fixed in public view. Darius's account described their claims and their defeat. Even if some supporters remained loyal in private, the official version stood high above the road. Families tied to those leaders carried the mark of association. In regions once ruled by the named rebels, local elites faced scrutiny from Persian officials.

Artisans and scribes who carved the inscription also worked under strain. A mistake in royal text cut into living rock could not be easily erased. Supervisors likely inspected each line before chisels moved on. The work took place high above the ground, where wind and height added danger. The relief showed bound figures beneath the king's foot. Those who carved it understood the message clearly. Their own safety depended on shaping the king's claim without error.

Records Copied Across Provinces

After the inscription was completed, Persian administrators ensured that its account of rebellion and victory was preserved in written form. Clay tablet copies discovered in Babylon show that the text was transcribed and stored in archives. Officials in different regions could consult these records when recounting the king's rise. The cliff remained fixed, but its message moved through administrative channels.

There is no evidence that later Achaemenid rulers altered the original carving. The relief and text remained intact through the reigns that followed. Road maintenance crews kept the route below in use, while the inscription stood beyond easy reach. Over time, officials may have referred less often to the cliff itself, relying instead on archival copies. The carved stone remained in place, unchanged as imperial administration continued around it.

A Record Above Passing Empires

Long after Darius died, the Behistun inscription remained carved into the cliff. Successors ruled from Persepolis and Susa, yet the relief continued to overlook the road. Caravans from new provinces passed beneath it. Even when the Achaemenid Empire fell to Alexander in the fourth century BCE, the carving was not erased. The stone endured through shifts in power that swept across the region.

Over centuries, the languages carved into the cliff faded from common understanding. Travelers saw the figures but could no longer read the lines below them. Local stories formed around the relief, sometimes linking it to legends rather than Darius. The official account remained in stone, yet its meaning grew uncertain. The king's claim to rule by divine favor stood high on the rock, even as memory of the events it described began to blur.

Letters Cut Back into Meaning

In the nineteenth century, European travelers recorded the Behistun relief and attempted to copy its inscriptions. The height of the carving made access difficult. In the 1830s and 1840s, Henry Rawlinson climbed the cliff and traced the Old Persian text by hand. Later scholars compared these copies with Babylonian and Elamite versions preserved on the rock. The parallel texts became key to unlocking cuneiform script.

Today, the relief still rises above the road in western Iran. Modern surveys document tool marks, weathering, and cracks in the limestone. Protective measures limit damage from pollution and erosion. Visitors look up at the carved king and bound figures much as ancient travelers once did. The cliff holds the names of rebels and the claim of a ruler who set his story into stone, high above the path of passing empires.

Chapter 10
The Shield That Set the Limits — The Twelve Tables and Roman Civil Authority

In the middle of the fifth century BCE, citizens gather in the Roman Forum beneath a bright sky. Dust rises around their sandals as magistrates step forward carrying wooden boards coated in white paint. The boards are heavy, their surfaces freshly marked with dark letters. Patricians in bordered togas stand apart, arms folded. Plebeians crowd closer, faces tense with expectation.

For years, legal judgments have been shaped by unwritten custom, known mainly to elite families. Disputes over debt, land, and injury have stirred anger across the city. Now the decemvirs, appointed to draft written law, raise the tablets into public view. A scribe begins to read aloud, his voice carrying over the murmur of the crowd. The rules are plain and direct.

They name penalties, procedures, and rights. In this moment, law moves from guarded memory to painted board, set before the eyes of Rome.

A Republic Divided by Status

In the early Roman Republic, power rested in the hands of patrician families who controlled key offices and claimed deep knowledge of ancestral custom. Plebeians, who made up much of the army and labor force, had fewer rights in political decision making. Disputes over debt and land often ended in favor of those who could interpret the law. Tension rose as plebeian citizens demanded clearer protection against arbitrary judgment.

The conflict led to the appointment of a board of ten men, known as the decemvirs, tasked with drafting written laws. For a time, normal magistracies were suspended so this board could work without interference. Envoys were said to have studied Greek legal practices, though Roman tradition shapes much of this account. In the Forum, citizens waited as the decemvirs prepared to present rules that would bind both classes under public authority.

Laws Painted for Public View

According to Roman tradition, the first set of laws was drafted in 451 BCE and displayed on ten wooden tablets. The following year, two more were added, bringing the total to twelve. The text was written in concise Latin, using clear phrasing meant to be recited aloud. Scribes painted the letters in dark pigment against a white coated surface so they could be read in the open air of the Forum.

The tablets were set up in a public space where citizens could see them. The boards were not carved in stone at first, but made from wood prepared for display. Over time, the original tablets were lost, likely destroyed during the Gallic sack of Rome in 390 BCE. Later generations preserved their content through memorization and citation in legal practice. The laws continued to circulate in written and spoken form.

Citizens Listening in the Forum

When the tablets were raised in the Forum, citizens gathered to hear them read aloud. Plebeians pressed forward, eager to learn rules that would limit the power of magistrates. Patricians stood nearby, aware that written law would reduce their exclusive control over custom. The language was plain, setting out procedures for court summons, debt settlement, property lines, and family authority. Many in the crowd nodded as familiar disputes were named in clear terms.

Yet not every line brought comfort. Some provisions confirmed harsh penalties for unpaid debt and strict control within households. Plebeians listened closely, weighing gains against continued burdens. Patricians watched for any clause that might weaken their status. The boards stood in open view, but uncertainty remained about how these painted words would shape daily judgment in Rome's courts.

Tension Within the Decemvirate

As the Twelve Tables took effect, pressure mounted within Rome's leadership. The decemvirs held wide authority during their term, replacing regular magistrates. Some citizens began to question whether the board would surrender power once its work was complete. Rumors spread through the Forum that certain members favored extending their control. Patrician families watched carefully, concerned about any shift in balance within the Senate.

Religious officials also felt strain. The laws touched on burial practices and sacred boundaries, areas long guided by priestly oversight. If written statutes contradicted ritual custom, conflict could follow. Plebeian leaders pressed for enforcement that favored fairness, while conservative voices urged caution. The tablets, meant to settle dispute, instead sharpened attention on who would interpret and apply them in daily governance.

Learning the Words by Heart

Roman citizens were expected to know the contents of the Twelve Tables. Later writers report that schoolboys memorized portions of the laws as part of their education. While the original wooden tablets did not survive, references in authors such as

Cicero and Livy preserve fragments of their wording. Magistrates consulted these lines when hearing cases in the Forum.

Interpretation quickly became a matter of skill. The laws were brief, leaving room for debate over meaning. Patrician jurists began to shape early legal commentary, guiding how each clause applied in practice. There is no surviving record of a formal commentary from the fifth century BCE, but later legal traditions trace their authority back to these texts. Control over interpretation remained a source of influence, even after the rules were publicly displayed.

Names Called Before the Court

For ordinary Romans, the Twelve Tables brought disputes into sharper light. A creditor could summon a debtor according to written procedure, and neighbors could point to fixed rules in arguments over land. Public knowledge of the law meant that mistakes or failures became more visible. A man judged in the Forum stood before clear penalties that others could quote back to him.

For members of the decemvirate, the cost was political. Later Roman tradition tells of Appius Claudius, whose conduct during his term stirred anger among plebeians. Though details are shaped by later writers, the story reflects strain tied to concentrated power. Officials who interpreted the tablets faced scrutiny from both classes. The painted boards did not remove conflict. They placed names and actions within a framework that citizens could see and remember.

Law Folded into Daily Governance

After the decemvirs' term ended, regular magistracies resumed, and the Twelve Tables remained part of Rome's legal foundation. Praetors and other officials applied the written clauses in court, citing them during hearings. Over time, additional statutes and edicts expanded Roman law, yet the Tables were still treated as a base text. Legal procedure grew more complex, but references to the early code continued.

The original wooden boards did not survive the Gallic sack of Rome, according to later accounts. What endured were recitations and written citations embedded in legal practice. Jurists preserved phrases and examples in teaching and argument. The absence of the physical tablets did not erase their authority. Instead, the content was absorbed into Rome's expanding administrative system, carried forward through repeated use in court and civic life.

A Code Remembered Through Centuries

Generations after the Twelve Tables were first displayed, Roman writers still referred to them as the root of civic order. Cicero, writing in the first century BCE, praised their clarity and authority. Though Rome had grown from a city state into a vast republic and later an empire, jurists traced many legal principles back to those early boards. The idea that law should be written and public endured as Rome expanded.

Yet uncertainty lingered around the exact wording of some clauses. Because the original tablets were lost, later copies relied on memory and citation. Scholars debated how closely surviving fragments matched the first display in the Forum. As new laws accumulated, the Twelve Tables became both foundation and symbol. Their absence as physical objects left space for interpretation, even as their legacy shaped Roman civil authority for centuries.

Fragments Carried in Later Texts

No physical tablet from the fifth century BCE has survived. What remains are quotations preserved in later Roman literature and legal commentary. Authors such as Cicero, Gaius, and Aulus Gellius cited portions of the Twelve Tables in their works. These fragments allow modern scholars to reconstruct parts of the code, though gaps remain. The laws live on through lines embedded in speeches and legal manuals.

Archaeological excavations in the Roman Forum have uncovered layers of paving and foundation from early civic spaces, but not the original boards themselves. Museums display inscriptions from later Roman law carved in stone, showing how written statutes became permanent fixtures in public life. The Twelve Tables endure as words carried forward in text, their painted boards long vanished from the square where citizens once gathered to hear them read aloud.

Chapter 11
The Spear That Chose a King — The Frankish Lance and Royal Acclamation

Cold air hangs over a wooden hall in early medieval Gaul. Torches hiss against damp stone as warriors press shoulder to shoulder beneath heavy beams darkened by smoke. At the center of the hall, a tall spear rests upright, its iron head catching firelight. The shaft is worn smooth where hands have gripped it over years of campaign. A circle opens around it as a young noble steps forward.

Outside, horses stamp in frost. Inside, murmurs rise and fall like wind through trees. The assembled warriors lift the spear and raise it above the young man's head. Shields strike against wood and stone in rhythmic blows. The sound fills the hall. No crown rests on his brow, no throne beneath him. The spear held high serves as the sign. In that lifted moment, authority passes not through parchment or seal, but through iron and the acclaim of armed men.

Rule Among Warrior Bands

In the fifth and sixth centuries, Frankish leaders ruled through networks of loyalty tied to warfare and land. Authority rested not only in bloodline, but in the support of armed followers. After the fall of Roman control in Gaul, regional strongmen competed for power. Clovis, king of the Salian Franks, expanded his rule through battle and alliance, setting a pattern for later succession.

When a king died, his realm did not pass smoothly to a single heir without contest. Sons, brothers, and rival nobles gathered support among warriors. Assemblies of armed men met in halls or open fields to declare their choice. Ritual gestures carried weight in these gatherings. A raised spear or shield could signal acceptance. In this setting, physical objects in the hands of warriors marked the transfer of authority before any written charter confirmed it.

Iron Forged for War and Symbol

The lance used in Frankish acclamation began as a weapon shaped by a blacksmith's skill. Iron was heated in a charcoal furnace, hammered into a narrow blade, and fitted onto a wooden shaft. The head was designed for thrusting in close combat, balanced for strength rather than ornament. The shaft was cut from straight grained wood, trimmed and smoothed for grip.

Over time, certain spears gained status beyond their use in battle. Chroniclers such as Gregory of Tours describe rituals in which a spear was raised over a chosen leader. The object itself was not carved with long inscriptions or adorned with jewels. Its authority came from shared recognition among warriors. What began as a practical tool of war became a visible sign in assemblies, linking the act of acclamation to the weapon carried into combat.

Eyes on the Raised Weapon

When a spear was lifted over a candidate in assembly, warriors watched closely for signs of support. Some struck their shields in approval. Others stood silent, weighing the strength and character of the man before them. The gesture was simple, yet it demanded agreement from those who would fight under his command. The raised lance signaled intent, but it did not erase doubt.

Clerics present at later gatherings observed the ritual with mixed feelings. As Christianity spread among the Franks, bishops sought to guide succession through blessing and prayer. The sight of authority passing through a weapon reflected older custom rooted in warrior culture. Some church leaders accepted the act as tradition, while others urged a stronger role for sacred ceremony. The hall held both expectation and hesitation as iron gleamed above a chosen head.

Between Sword and Altar

As Frankish kingdoms expanded, succession grew more complex. Rival heirs gathered support among regional nobles, and assemblies could divide along family lines. The act of raising a spear over one claimant risked angering another faction. Warriors pledged loyalty in public view, knowing that refusal might be remembered. Tension filled halls where allegiance was declared by weapon and voice.

Church leaders sought a stronger presence in royal selection. Bishops urged that kings receive formal blessing and anointing, tying rule to Christian ritual. Some nobles supported this shift, seeing in it a way to limit open rivalry. Others held fast to older forms of acclamation that placed decision in the hands of armed men. The spear remained central in assemblies, yet pressure grew to bind its meaning to sacred authority as well.

Chroniclers and Changing Ritual

Our knowledge of Frankish acclamation comes from written accounts such as Gregory of Tours and later Carolingian sources. These texts describe warriors raising a leader on a shield or lifting a spear to signal approval. The records do not preserve a fixed script or formal wording tied to the act. Instead, they note the gesture itself as a visible sign of acceptance among armed followers.

As royal ceremony developed, written capitularies and church councils sought to define clearer procedures for succession. Anointing with holy oil began to accompany or replace older warrior rites. Surviving charters from later periods emphasize blessing and coronation rather than weapon lifting. Some later traditions romanticized the spear as sacred or miraculous, but contemporary sources focus on its role in public acclaim. The meaning of the gesture shifted as institutions worked to frame it within new ritual order.

Loyalty Measured in Public View

For a noble whose spear was raised in support of a claimant, the choice carried risk. If the chosen leader secured the throne, loyalty could bring land, rank, or favor. If he failed, those who had stood beneath his banner faced loss or exile. Public acclamation left no room for quiet neutrality. The act of lifting a weapon marked allegiance in front of peers who would remember.

For the chosen king, the moment of acclaim brought its own burden. Warriors expected reward, justice, and victory in battle. A ruler who failed to protect his followers or divide spoils fairly could see support weaken. The same hall that echoed with shield blows could fall silent at the next gathering. Authority gained through iron and voice rested on continued trust among men accustomed to contest and war.

From Assembly to Coronation

By the eighth century, Frankish rulers began to formalize succession with greater involvement from the church. The anointing of Pepin the Short in 751 marked a shift toward ceremony led by bishops. Written records describe blessing and oath taking alongside public acclaim. The spear did not disappear from royal imagery, but its role became one element within a broader ritual framework.

Royal capitularies and church councils addressed order in succession and loyalty. Oaths were recorded, and assemblies were guided by established procedure. Over time, coronation with crown and sacred oil gained prominence in written accounts. The earlier practice of raising a weapon remained part of memory and narrative, yet official documentation focused increasingly on liturgy and oath. Authority moved from the open hall of warriors into structured ceremony shaped by both noble and church.

Memory of the Raised Spear

Centuries after early Frankish assemblies, chroniclers looked back on the act of weapon acclamation as part of a heroic past. Carolingian writers described earlier kings lifted before their warriors, linking new rulers to long lines of martial leadership. The memory of the spear in the hall reinforced the idea that royal authority rested on the consent of armed followers, even as coronation rituals grew more elaborate.

Yet uncertainty lingered in times of succession crisis. When rival claimants emerged in later centuries, nobles still gathered in assembly to declare support. Though crowns and anointing oil marked kingship, the backing of powerful warriors remained decisive. The raised weapon, whether literal or recalled in story, stood as reminder that rule depended on visible acclaim. Even as institutions strengthened, the echo of shield blows in timber halls continued to shape expectations of power.

Weapons in Cases of Glass

No single spear from an early Frankish acclamation can be identified with certainty. The weapons used in assemblies were ordinary tools of war, subject to wear and loss. Archaeologists have recovered Frankish lance heads from graves and settlement sites, showing iron blades shaped for thrusting and mounted on wooden shafts long since decayed. These objects offer material proof of the weapons carried by the warriors who gathered in halls.

Written sources preserve the memory of the ritual more clearly than any surviving spear. Chronicles copied in medieval scriptoria recount the raising of kings by armed men. Today, museums display lance heads behind glass, labeled by date and region. The iron is dark with age, edges worn by time. Though the specific spear that marked a king is gone, the practice remains recorded in metal fragments and in ink that outlived the timber halls.

Chapter 12
The Crown That Bound the Realm — The Hungarian Holy Crown and Legal Kingship

Candles tremble in the cold air of the cathedral, their flames bending as drafts slip through stone arches. Nobles in fur lined cloaks stand shoulder to shoulder beneath painted vaults darkened by smoke. At the center of the nave, a cushion of deep red velvet rests on a narrow table. Upon it lies a crown of gold plates and enamel images, its surface glowing in shifting light. A small cross rises from the top, tilted slightly to one side.

The archbishop lifts the crown with both hands. It is heavier than it appears. The metal presses cold against his palms as he turns toward the waiting ruler. Armor creaks softly as knights shift their stance. No trumpet sounds. No cheer breaks the silence. All eyes follow the movement of the crown as it lowers toward a bowed head.

When the gold touches the brow, the moment fixes in place. Without this crown, lineage alone would not suffice. With it, kingship becomes lawful. In this hushed space of stone and flame, an object completes the act of rule.

A Kingdom Forged in Faith and Law

In the late tenth and early eleventh centuries, the lands of the Carpathian Basin were still shaping their identity. Tribal leaders who had once moved across open plains now faced pressure from powerful Christian neighbors. Stephen, son of Géza, accepted baptism and sought recognition from Rome. When he was crowned around the year 1000, Hungary entered the circle of European kingdoms bound by Christian law and church authority.

From that point forward, rule in Hungary did not rest on bloodline alone. Bishops, abbots, and leading nobles formed a network that balanced royal command with established custom. Laws were issued in the king's name, but they were enforced through counties, monasteries, and royal officials. Over time, a clear understanding took hold. A ruler was not fully king unless he was crowned in the proper place, by the proper church authority, with the proper crown.

Gold, Enamel, and a Crown Assembled

The crown now known as the Holy Crown of Hungary was not made in a single workshop at one moment in time. Its lower band, called the corona graeca, bears Greek inscriptions and enamel portraits of Byzantine figures, including Emperor Michael VII Doukas. Scholars date this portion to the late eleventh century. The upper cross bands, known as the corona latina, carry Latin inscriptions and images of apostles, added separately and joined to the lower ring.

Gold plates were shaped, pierced, and riveted together by skilled metalworkers. Enamel plaques were fired in small furnaces, their colored glass fused into carved recesses. Tiny pins and hinges fixed each image in place. The cross at the top was mounted later and now tilts from damage sustained centuries ago. The crown's current form reflects careful assembly from distinct parts, shaped by artisans who worked under royal and ecclesiastical oversight.

A Crown Watched by Lords and Bishops

When the crown first appeared in full ceremonial use, nobles and bishops watched it closely. Its Byzantine images drew notice. Greek lettering circled the lower band, unfamiliar to some Hungarian lords who spoke Latin in church and court. Clergy studied the plaques and spoke in low voices about their origin. The presence of foreign craftsmanship did not pass without comment. It raised quiet questions about alliances and influence beyond Hungary's borders.

Yet hesitation did not turn into open resistance. The archbishop accepted the crown as fit for consecration. Leading magnates measured its weight and form with careful eyes. What mattered most was not where each plate had been forged, but whether this crown would become the fixed instrument of lawful rule. In corridors and chapter houses, men weighed the risk of division against the need for unity under a recognized, visible sign of kingship.

Rival Claims Beneath a Single Crown

As decades passed, the crown drew more than reverence. It drew competition. When a king died without a clear heir, rival claimants moved quickly to secure backing from powerful nobles. Yet support in arms was not enough. Bishops insisted that no ruler could be fully lawful without coronation using the recognized crown. This demand placed the object at the center of political struggle.

Whispers spread through court circles when the crown was removed from one stronghold to another. If it rested in the custody of one faction, others feared exclusion. Some lords pressed for swift coronation before opposition could gather. Church officials worried about improper rites or unauthorized handling. The crown's movement across roads and rivers carried tension with it. Its absence from the cathedral treasury stirred quiet alarm among those who believed the realm's stability depended on its controlled use.

Guarded Access and Careful Custody

Royal and church records show that the crown was placed under strict supervision. It was kept in guarded treasuries, often within fortified castles or cathedral vaults. Selected nobles and clerics were assigned as custodians. Written charters recorded who held responsibility for its safekeeping. When coronations approached, formal instructions governed who could carry it, who could touch it, and where the ceremony must occur.

Later stories claimed secret trials or hidden oaths tied to the crown, but surviving medieval documents focus instead on procedure. Disputes over its custody appear in legal records, especially during contested reigns. Councils met to confirm that the proper rites had been followed. At times, officials ordered inventories of royal regalia to ensure that no piece had been altered or removed. These efforts did not center on symbolism alone. They aimed to control the physical object that defined lawful kingship.

Men Marked by Custody and Oath

Those entrusted with the crown carried more than ceremonial duty. If the crown was lost, delayed, or seized, suspicion fell first on its guardians. Noble families tied to its custody risked their standing at court. In times of conflict, accusations spread quickly. A lord who allowed the crown to pass into rival hands could face charges of betrayal, even if force had been used against him.

Clerics also bore strain. An archbishop who crowned the wrong claimant faced exile or loss of office if power shifted. Some churchmen hesitated before performing coronations during disputed successions, knowing their choice could divide the realm. Private letters and later chronicles record tension within cathedral chapters when pressure mounted. Serving the crown meant serving stability, yet it also meant standing at the edge of danger when loyalties fractured.

Rules Written Around a Sacred Object

As succession crises repeated across centuries, formal rules hardened around the crown. Legal custom required coronation in Székesfehérvár, performed by the archbishop of Esztergom, using the recognized crown. Royal decrees and later legal texts repeated these conditions. If any element was missing, nobles could challenge the ruler's status. The process became fixed in law, not left to memory.

When the crown had to be moved, written records tracked its transfer. Guards were appointed in set numbers. Seals were placed on chests and doors. During foreign invasions or civil unrest, officials ordered the crown hidden in strongholds or carried to secure locations. At times it crossed borders for safekeeping, escorted by nobles under oath. Administrative silence followed each transfer, recorded briefly in charters without ornament, as though the object's movement required careful restraint rather than public display.

A Crown That Outlasted Dynasties

Over the centuries, dynasties rose and fell, yet the crown remained the fixed measure of lawful rule. In the fourteenth and fifteenth centuries, rival kings who gained military control still faced challenge if they lacked proper coronation with the crown. Legal scholars wrote that sovereignty rested not in the person alone, but in the Crown itself. This idea shaped Hungary's political language for generations.

Even in later periods of foreign domination, debates continued over where the crown should rest and who had the right to guard it. When it was removed during war, anxiety followed. Letters and council records reveal concern that the absence of the crown weakened the realm's standing. Though the monarchy endured in changing form, uncertainty lingered. As long as the crown's custody shifted under pressure, questions about rightful authority did not fully disappear.

Glass, Steel, and a Crown Under Watch

Today the Holy Crown rests under guarded display in the Hungarian Parliament Building in Budapest.

Climate controls regulate temperature and humidity. Security officers stand at measured distance. Visitors move past in quiet lines, pausing to study the gold bands and enamel plaques. The tilted cross remains visible, its angle preserved rather than corrected. The crown's surface shows wear from centuries of handling, transport, and ceremony.

Archival records survive in royal charters, cathedral inventories, and later legal writings that defined its role. Scholars examine the inscriptions in Greek and Latin, the metal composition, and the joinery of its parts. No medieval workshop record explains every stage of its creation, and debates continue about how the components were first combined. Beneath museum lights, the crown's enamel faces reflect back faint images of those who look at it, an object still bound to law and memory.

Chapter 13
The Drum That Summoned the Court — The Zhou Dynastic Bells and Ritual Authority

Rain has just passed over the walled city, leaving the packed earth dark and heavy underfoot. Inside a broad courtyard, workers brace their shoulders against thick wooden beams as they lift bronze bells into place. Each bell is large and oval, its surface cast with raised patterns and lines of inscription near the lip. Ropes strain. Wood creaks. When a bell settles onto its carved frame, the sound of metal against wood carries in a low, dull note.

Officials in layered robes stand beneath the eaves, watching the arrangement take shape. Servants wipe moisture from the bronze, revealing a faint golden sheen beneath the gray surface. Musicians test a strike with a wooden mallet. The bell answers with a deep, measured tone that rolls across the courtyard walls and out toward the city beyond.

The sound is not casual music. It marks rank and order. It signals that the ruler presides in harmony with Heaven. As more bells are raised and aligned in careful rows, the court prepares to let bronze speak for authority.

A Court Governed by Ritual and Rank

During the Eastern Zhou period, royal power no longer flowed smoothly from the central king. Regional states grew stronger, and rulers guarded their own territories with armed retainers and fortified cities. Yet even as political control fractured, shared ritual forms continued to bind elite society together. Courts across the region observed formal ceremonies that affirmed hierarchy, loyalty, and inherited rank.

Music stood at the center of these rites. Officials recorded titles, duties, and noble lineages with care. A ruler's standing could be measured by the size and number of ritual vessels and instruments he possessed. Bronze bells were not simple decorations for celebration. They were assigned by rank, regulated by court custom, and displayed during state ceremonies. When a lord gathered his ministers and struck the bells in proper order, he signaled not only wealth but recognized authority within a system shaped by long standing ritual law.

Bronze Cast in Fire and Mold

The bells known today as the Bianzhong of Marquis Yi of Zeng were cast in the fifth century BCE. Craftsmen prepared clay molds in sections, shaping the inner core and outer shell with precise spacing. Molten bronze, heated in large furnaces, was poured between these layers. The alloy mixture of copper, tin, and small trace elements was measured with care to control hardness and tone.

Once cooled, the molds were broken away, and the surface was filed and polished. Artisans carved inscriptions into the bronze, recording the name and rank of the marquis and noting the musical scale of each bell. Small raised bosses were cast along the sides to guide tuning. Each bell could produce two distinct tones depending on where it was struck. The finished instruments were hung on wooden frames built to hold them in ordered rows.

Courtiers Listening for Meaning

When the full set of bells was first arranged in the court of Marquis Yi, ministers gathered to witness the display. They stepped closer to read the inscriptions cast into the bronze. The characters named the marquis and marked his authority within a ranked system of states. Musicians tested each tone while scribes checked the sequence against known pitch standards. The sound was strong and steady, but some officials listened with narrowed eyes.

Regional rivals had long measured status through ritual vessels. A bell set of this size signaled ambition. Envoys from neighboring states noted the number of instruments and the precision of their tuning. Quiet concern moved through the ranks of lesser lords who lacked such display. The bells did not speak in words alone. Their scale and sound suggested a claim to standing that others could not ignore.

Ritual Rank Under Watchful Eyes

As news of the bell set spread, larger states took notice. Messengers carried reports of the number of bells and the complexity of their tones. In the Zhou ritual order, musical instruments were tied to rank. Only certain rulers were entitled to possess large, tuned sets. If a regional lord appeared to exceed his station, questions followed. Court advisors in neighboring capitals compared the display to long standing ritual codes.

Within the marquis's own court, senior officials debated the limits of proper ceremony. Some argued that the bells affirmed loyalty to the Zhou king through shared ritual form. Others worried that excess display could invite scrutiny or resentment. Priests responsible for sacrificial rites reviewed the inscriptions to ensure that titles and honors matched recognized status. The bells stood silent between performances, yet the pressure around their presence continued to build.

Measured Tones and Recorded Authority

Court scribes copied the inscriptions from the bells onto bamboo slips for archival record. They listed the titles carved into the bronze and compared them with existing registers of rank. Ritual masters supervised rehearsals, striking each bell at marked points to confirm proper pitch. The placement of the instruments followed written guidelines preserved in court manuals, though some details of those manuals survive only in later texts.

Centuries afterward, classical scholars claimed that Zhou rulers enforced strict musical laws across all states, but surviving evidence shows uneven control. Archaeological finds reveal variation in bell numbers and tuning systems. In the marquis's court, officials limited who could perform and who could approach the frames. Access to the bells was restricted during non ceremonial days. Their inscriptions were not altered, and no record suggests public recarving, but careful oversight guarded how and when their sound could shape authority.

Musicians and Lords Under Scrutiny

The musicians chosen to strike the bells carried heavy responsibility. A missed tone or broken rhythm during a formal rite could bring public shame. Court records from later periods describe punishments for ritual errors, and similar pressure likely weighed on performers in the marquis's hall. Young players trained for years, yet their hands still trembled when nobles filled the courtyard and foreign envoys stood watching.

The marquis himself bore quiet strain. By displaying a bell set of this scale, he raised expectations among his peers. If he failed to maintain proper order in his territory, rivals could point to the bells as proof of overreach. Advisors urged caution in hosting large ceremonies too often. The bronze instruments, once cast and hung, could not be hidden easily. Their presence tied the marquis's reputation to every sound that echoed across the court.

Ritual Codes Written in Practice

Over time, court officials established fixed procedures for storing and maintaining the bells. When not in use, the instruments were covered with cloth and guarded within enclosed halls. Wooden frames were inspected for weakness, and bronze surfaces were wiped to prevent corrosion. Written tallies listed the number of bells and their assigned pitches. Access required approval from senior ritual officers.

Ceremonial schedules were recorded in advance. Only certain state events permitted full performance of the set. Lesser gatherings called for fewer instruments, in keeping with rank. Envoys from other states were received under carefully planned rites that displayed authority without exceeding accepted limits.

Surviving ritual texts from the Zhou tradition describe similar controls in other courts, suggesting that formal oversight shaped how and when such instruments could be heard. The bells' sound followed regulated order, guided by recorded custom rather than impulse.

A Sound That Outlived the Court

After the death of Marquis Yi, the bell set was buried with him in a sealed tomb. The instruments did not pass to a successor's hall. Instead, they were lowered into the earth along with chariots, weapons, and attendants. The decision removed the bells from active ritual life. Their tones fell silent above ground, and the political order they once reinforced shifted under new rulers.

In the decades that followed, regional states continued to compete for status through ritual display. Yet the memory of large, tuned bell sets endured in written tradition. Later texts spoke of ideal music ordered under Heaven, though they did not describe every court in equal detail. Uncertainty remained about how closely real practice matched recorded theory. The bronze lay underground, preserving a moment of authority even as the world above it changed course.

Bronze Unearthed Beneath the Soil

In 1978, workers in Hubei Province uncovered a sealed tomb complex at Leigudun. Inside lay the intact bell set of Marquis Yi of Zeng, arranged on wooden frames that had partly decayed but still held their shape. The bronze surfaces remained remarkably preserved. Inscriptions were legible, confirming titles, ranks, and pitch markings. Archaeologists documented each bell's position before removal, mapping their order with measured care.

Modern researchers tested the tones in controlled settings, striking the bells with padded mallets. Each instrument produced two distinct notes, just as the inscriptions indicated. Conservation teams stabilized the metal and reconstructed the wooden frames for display. Today the bells stand in museum halls, suspended in ordered rows. Their bronze faces reflect soft light, and when replicas are sounded, the deep tones roll outward once more, echoing a court long buried beneath the earth.

Chapter 14
The Banner That Defined a People — The Aztec Codex Mendoza and Imperial Tribute

In a stone chamber in Mexico City in the 1540s, light falls through a high window onto a long wooden table. Sheets of pale European paper lie flat, weighted at the corners by smooth stones. Indigenous painters sit side by side, their brushes tipped with black and red pigment. One grinds color in a small bowl. Another draws a town glyph with a steady hand, outlining a temple and a hill.

Across the table, Spanish officials lean forward, watching closely. They do not speak Nahuatl fluently, yet they follow the images. Bundles of cloth appear in neat rows. Shields stack beside baskets of maize. Feathers, cacao beans, and jaguar skins fill the page in careful order. Each image marks a place and the goods it must deliver.

Outside, the great city once called Tenochtitlan still rises from the lake, though its rulers have changed. Inside this room, tribute is being fixed onto paper. The brush moves without pause, recording a system that once fed an empire and now stands open for new masters to study.

An Empire Measured in Tribute

Before the Spanish conquest, the Mexica ruled through a network known as the Triple Alliance. Tenochtitlan, Texcoco, and Tlacopan formed the core of this power. Conquered towns were not always stripped of local leaders, but they were bound by tribute. Officials recorded which provinces delivered cotton cloth, maize, cacao, feathers, rubber, and warriors for campaign. Goods moved along causeways and canals toward the island capital.

Tribute collectors operated under strict schedules. Delegations arrived at set intervals, bringing counted bundles and measured goods. Imperial overseers verified deliveries in public spaces, often before temple precincts or palace courts. Painted records, kept by trained tlacuilos, showed place glyphs beside tribute items. These images were not decoration. They marked obligation in a system that tied economic flow to military authority and reinforced the central rule of Tenochtitlan.

Paper, Pigment, and Ordered Pages

The Codex Mendoza was created around 1541, likely under the direction of Viceroy Antonio de Mendoza. Indigenous painters worked on European paper rather than traditional bark or deer hide. The sheets were bound along one edge to form a single volume. Black ink outlined figures and place glyphs, while red, blue, yellow, and green pigments filled in details with careful strokes.

The painters followed established pictorial conventions used before the conquest. Town symbols were drawn first, then tribute goods were placed in rows beside them. Spanish scribes later added alphabetic notes in the margins, explaining the images in Castilian. The work required collaboration. Indigenous artists shaped the visual record from memory and prior models. Spanish officials oversaw the process, ensuring the tribute lists matched what colonial authorities expected to see.

Officials Reading Painted Records

When the codex pages began to fill, Spanish administrators examined them closely. They traced the rows of tribute goods with their fingers, counting bundles and noting town names written in the margins. Some nodded in approval at the clear layout. Others asked questions through interpreters, unsure whether each painted symbol matched current obligations. The images revealed a structure of rule that had once operated with steady precision.

Indigenous nobles also viewed the pages with guarded expressions. They recognized the place glyphs and the tribute tallies from earlier records. Yet now these details were being compiled for foreign rulers. A few hesitated before confirming certain counts, aware that mistakes could carry consequences. The codex did not merely preserve memory. It translated an imperial system into a form that new authorities could inspect and use.

A Ledger That Raised Questions

As more pages were completed, tension grew around the table. Spanish officials compared the tribute lists in the codex with reports from colonial tax collectors. If painted records showed higher past deliveries, questions followed. Were current towns withholding goods, or had tribute been reduced under new rule? Administrators pressed interpreters for clarity. The codex began to serve as a standard against which present demands could be measured.

Among Indigenous leaders, quiet concern spread. The images fixed earlier tribute levels in lasting form. Some feared that the book might be used to restore heavy obligations or justify stricter enforcement. Franciscan friars, focused on conversion and reform, watched the project with interest. They saw in the codex a record of former power structures tied to temples and war. As pigment dried on each page, the painted tribute system drew scrutiny from more than one direction.

Translating Empire Into Margins

Spanish clerks added alphabetic notes beside the painted images, explaining each province and its tribute in Castilian. Interpreters stood between painters and officials, clarifying the meaning of glyphs and quantities. The book was reviewed page by page before being prepared for shipment to Spain. Surviving copies show corrections in ink, where a symbol was adjusted or a gloss refined.

Later writers claimed that the codex was meant to glorify Aztec rulers for European readers, but contemporary evidence points to a practical purpose. The volume functioned as a report for royal authorities overseas. Access was limited once completed. It was packed carefully and entrusted to a ship bound for Spain. The codex left New Spain under official oversight, intended for inspection by those who governed from across the ocean.

Artists Between Two Worlds

The Indigenous painters who worked on the codex carried quiet strain. They drew the tribute system of their former rulers while answering to Spanish supervisors. Some had family ties to noble houses that once benefited from the flow of goods into Tenochtitlan. Now they recorded those same obligations for a colonial regime. Their skill secured them work, yet it also tied them to the reshaping of their own political past.

Spanish officials faced pressure as well. If the codex failed to reflect accurate tribute levels, blame could fall on those who oversaw its preparation. Letters from colonial administrators show concern about reporting errors to the Crown. Misrepresentation could damage reputations or careers. Around the long table, brush strokes and margin notes carried more than pigment and ink. They carried risk for the men who stood behind them.

A Manuscript in Transit and Custody

After completion, the codex was prepared for shipment to Spain as part of official correspondence. It was folded, packed, and sealed among other documents bound for the royal court. During its journey across the Atlantic, the vessel was intercepted by French privateers. The book changed hands before ever reaching its intended audience. From there, it entered European collections, moving through private custody rather than royal archive.

In later decades, the codex was preserved in controlled settings, cataloged alongside other manuscripts. Owners restricted access, allowing scholars to consult it under supervision. Marginal notes were examined but not erased. The pages were not altered to fit new policy. Instead, the manuscript was stored, transferred, and safeguarded through formal record. Its handling followed archival practice, shifting from colonial report to a guarded object within European libraries.

A Record That Outlived Conquest

In the decades after its creation, the codex became one of the clearest surviving records of the Aztec tribute system. As colonial administration evolved, Spanish authorities relied more on new tax structures than on older imperial patterns. Yet the painted lists remained a reference point for understanding how central Mexico had once been organized. Scholars and officials consulted it when questions arose about former provinces and obligations.

Centuries later, historians used the codex to reconstruct the reach of Mexica power before 1521. The images revealed the scale of goods once demanded and delivered. Still, uncertainty lingered. Not every province's tribute is fully preserved, and some pages show signs of wear and loss. Gaps in the record leave room for debate about regional variation and enforcement. The painted rows endure, but parts of the imperial system they depict remain partly obscured.

Pages Preserved Across Oceans

Today the Codex Mendoza is housed in the Bodleian Library at the University of Oxford. The paper shows signs of age, with slight discoloration along the edges and faint creases where pages were once folded. Conservators monitor humidity and light to prevent further fading of pigments. Each page has been digitized, allowing detailed study without repeated handling of the original sheets.

Scholars examine brush strokes, pigment composition, and the layering of Spanish glosses over Indigenous images. The painted tribute goods remain clear, stacked in careful rows beside place glyphs that still mark towns across central Mexico. Under controlled lighting, the colors hold their shape against pale paper. The book rests behind glass, its painted bundles and shields silent but precise, a record of tribute that once sustained an empire and now endures in ink and fiber.

Chapter 15
The Charter in Wax and Silk — The Magna Carta and Constrained Kingship

Wind moves across the meadow at Runnymede, flattening tall grass along the banks of the River Thames. It is June 1215. Armed barons stand in clusters beneath their banners, mail glinting beneath open cloaks. Across the field, King John's pavilion rises in guarded silence. Horses stamp the damp ground. No one mistakes this gathering for ceremony alone.

At a rough wooden table set between rival camps, royal clerks unroll a long sheet of parchment. The surface is crowded with tight lines of Latin script. There are no painted images, no gold leaf, no decoration. Only clauses, written close together, leaving little space for air between demands and promises.

A small dish of melted wax sits nearby. The king's great seal, heavy and engraved, rests in a cloth wrap. When the seal is pressed into warm wax and fixed to the charter, the impression forms deep and clear. Around the table, men lean forward to watch. Steel has forced this meeting, but ink and wax will carry its terms outward, beyond the meadow and into the realm.

A Realm Strained by War and Tax

In the early thirteenth century, King John ruled a kingdom under heavy pressure. Military losses in France had cost England much of its continental territory. Campaigns required money, and the king demanded scutage payments from barons who did not serve in person. Sheriffs enforced levies with a firm hand. Royal courts extended their reach into disputes once settled by local custom.

Tension rose between crown and nobility. Powerful lords resented sudden fines and the seizure of lands during disputes. The Church had already clashed with John over the appointment of the Archbishop of Canterbury, leading to interdict and excommunication. Though reconciliation followed, mistrust remained. By 1215, armed barons gathered in open defiance, demanding written limits on royal power. The kingdom's political order, shaped by feudal ties and sworn loyalty, stood on uncertain ground.

Parchment, Ink, and the King's Seal

Royal clerks drafted the charter in Latin, the language of government and law. They wrote on parchment made from treated animal skin, scraped smooth and cut into long sheets. The text was arranged in tight lines, leaving little blank space. Scribes worked quickly but with steady hands, copying agreed clauses into formal script. There was no single decorated original. Multiple copies were produced for distribution to cathedrals and officials across the realm.

Once written, each copy required authentication. Warm wax was prepared, and the king's great seal was pressed into it. The seal bore the image of the king enthroned on one side and mounted on horseback on the other. No handwritten signature marked the document. The deep imprint of the seal confirmed royal assent, fixing the charter as a binding instrument under English law.

Barons Reading Their Demands in Ink

When copies of the charter reached cathedrals and shire towns, barons gathered to hear its clauses read aloud. In great halls and chapter houses, clerks recited lines that limited certain taxes and promised lawful judgment. Some nobles listened with guarded approval. The words matched demands they had pressed at Runnymede. For the first time, those demands were set in durable form.

Royal officials reacted with less ease. Sheriffs and justices studied the clauses that restricted their authority. Questions arose about how strictly the king intended to follow these terms. Even among the barons, uncertainty lingered. Would the crown honor the charter once the armed camps dissolved? The ink had dried, and the seals hung firm, yet quiet concern moved through those who understood how quickly power could shift when force returned to the field.

A Kingdom Watching for Breach

As weeks passed, tension tightened rather than eased. Royal advisors urged the king to reconsider the concessions pressed upon him. Letters moved between England and Rome, and word spread that Pope Innocent III had been asked to judge the charter's legality. Church leaders weighed their position carefully, aware that open support for the barons could weaken royal authority, yet silence carried its own risk.

Across the countryside, baronial leaders monitored the king's actions. They formed committees to enforce compliance, naming twenty-five of their number to hold the crown to its word. This clause alone raised alarm within royal circles. To the king's supporters, it suggested divided sovereignty. To the barons, it served as insurance against retreat. Suspicion deepened. Each side watched for signs of delay or defiance, unsure how long written terms could restrain armed men.

Reading Clauses Line by Line

Royal clerks reviewed the charter soon after it was sealed. They compared its clauses to existing feudal customs and earlier charters issued by King Henry I. Some provisions matched established practice, while others appeared to narrow the king's discretion in collecting taxes and imposing fines. Copies were sent to sheriffs with instructions to observe the terms, though surviving letters show uneven enforcement.

Later tradition described the charter as a sweeping declaration of liberty, but the 1215 text focused on specific grievances. Legal officers debated how to apply its clauses in active disputes. When confusion arose, local courts sought guidance from royal justices. The text itself did not change in those early weeks, yet its meaning was contested in practice. The parchment lay fixed, while interpretation shifted within royal and baronial councils.

Men Caught Between Oath and Crown

Barons who had forced the charter now faced strain among their own followers. Knights and retainers expected clear gains from the agreement. When royal officers hesitated to comply with certain clauses, frustration grew within baronial ranks. Some lords feared that open conflict might resume, risking lands and families already exposed during earlier campaigns. Reputation became fragile. A baron seen as too willing to compromise could lose support. One who pressed too hard risked being branded a rebel.

Royal officials endured similar pressure. Sheriffs tasked with enforcing the king's will had to balance loyalty to the crown with the written terms of the charter. A misstep could lead to dismissal or seizure of property. Church leaders also felt strain. Bishops who witnessed the sealing were drawn into dispute, accused by one side of overreach and by the other of hesitation. Personal bonds fractured under the weight of competing oaths.

Reissued Under New Authority

After King John's death in 1216, royal guardians acting for the young Henry III reissued the charter with revisions. Certain controversial clauses were removed, including the enforcement committee of barons. The revised text was sealed again and distributed to key centers. A further reissue followed in 1217, accompanied by a separate Charter of the Forest that addressed grievances about royal hunting lands.

In 1225, once Henry III reached majority, the charter was confirmed again in exchange for taxation granted by the realm. This version became the foundation for later references. Copies were entered into administrative rolls and preserved in cathedral archives. The charter was not left as a single artifact from Runnymede. It became part of formal record, reissued, sealed, and circulated through established channels of royal governance.

A Text Invoked Across Generations

In the decades that followed, Magna Carta was cited in disputes between crown and subject. Petitioners referred to its clauses when challenging unlawful imprisonment or excessive fines. Royal judges acknowledged its authority in court proceedings. Though not always enforced in full, the charter remained a touchstone in moments of conflict. Copies were read aloud at public gatherings to remind officials of agreed limits.

Even so, tension persisted. Later kings tested the boundaries of its provisions, and parliaments demanded reaffirmation. Each confirmation raised quiet doubt about how secure its promises truly were. Legal scholars debated wording and scope, comparing surviving copies to ensure consistency. The charter endured in parchment and seal, yet its power depended on those willing to invoke it. Its clauses lingered in law, but their reach was never entirely free from challenge.

Parchment Preserved Behind Glass

Four original copies of the 1215 Magna Carta survive today, housed in institutions such as the British Library and the cathedrals of Lincoln and Salisbury. The parchment shows creases from early handling and slight darkening from age. The script remains dense and even, written in steady lines by royal clerks. The wax seals that once hung from these copies are mostly lost, though slits and tags mark where they were attached.

Conservators monitor light and humidity to protect the fragile skin. Scholars compare the wording of each surviving copy, noting small variations in spelling and spacing. Visitors move past the cases in quiet lines, reading translated excerpts mounted nearby. The ink has faded in places, yet the clauses remain legible. Beneath museum lighting, the parchment lies still, a sheet once carried across a tense meadow, now resting as evidence of a kingdom forced to bind its ruler in writing.

Chapter 16
The Codex That Divided a Faith — The Donation of Constantine and Papal Authority

In a dim chamber in medieval Rome, a parchment lies stretched across a wooden desk. Candles burn low, their smoke drifting toward a vaulted ceiling stained by years of soot. A scribe bends forward, copying lines of Latin in careful script. The words claim that Emperor Constantine once granted sweeping lands and authority to the bishop of Rome.

Clerics gather nearby, robes brushing against stone floors. One traces a line with his finger, lips moving silently as he reads. Another pauses at a phrase describing imperial power transferred to the pope. The document carries no imperial seal from late antiquity, yet its claims are bold and direct.

Outside the chamber, church bells mark the hour. Inside, the parchment becomes more than ink on skin. It offers written proof of authority that reaches back to the age of emperors. As the scribe finishes a line and sharpens his quill, the chamber holds a quiet tension. If accepted, these words will strengthen the hand of Rome. If questioned, they will stir unease across Christendom.

A Church Rising Amid Fading Empire

After the fall of the Western Roman Empire in the fifth century, Rome no longer housed an emperor. Germanic kings ruled much of Italy, and imperial authority shifted east to Constantinople. In this vacuum, the bishop of Rome gained influence. Popes negotiated with invading armies, managed church lands, and acted as civic leaders in a city stripped of imperial administration.

By the eighth century, new political pressures shaped the papacy's position. The Lombards threatened Rome, and Byzantine power in Italy weakened. Popes sought alliances with Frankish rulers for protection. Written documents carried weight in these negotiations. Charters, letters, and copied decrees moved between courts and monasteries. Within this setting, a text that claimed ancient imperial endorsement could serve as strong support for expanding papal authority in both spiritual and temporal matters.

Ink on Parchment, Authority in Script

The text known as the Donation of Constantine survives in medieval manuscript copies rather than in any late Roman original. Scholars trace its likely composition to the eighth century, written in Latin consistent with that period rather than with the fourth century. Monastic scriptoria prepared parchment sheets, smoothed and cut for copying. Scribes wrote in steady hands, following established forms for legal and ecclesiastical documents.

The document claimed to record a decree issued by Emperor Constantine after his conversion to Christianity. It described the transfer of imperial regalia and authority over Rome and western territories to the pope. No imperial seal from Constantine's era appears in surviving copies. Instead, the text was transmitted through collections of canon law, copied alongside other church decrees. The material form reflects monastic production rather than Roman imperial chancery practice.

A Claim Read in Quiet Rooms

When the text began circulating within church collections, clerics read it alongside other legal authorities. Bishops and abbots copied its clauses into canon law compilations. The claim that Constantine had granted lands and symbols of imperial power to the pope drew attention. In some courts, the document was cited to support papal independence from secular rulers.

Yet hesitation appeared in certain circles. No older Roman copy could be produced. Some scholars noted differences in style between this text and known imperial decrees. These doubts did not lead to open rejection, but they lingered in private discussion. As the document moved through scriptoria and councils, it gained weight through repetition, even as questions about its origin remained beneath the surface.

A Text in the Midst of Power Struggle

By the eleventh century, tensions between popes and secular rulers sharpened during the Investiture Controversy. When disputes arose over who had the right to appoint bishops, supporters of papal authority pointed to written claims that reached back to Constantine. The Donation appeared in arguments that placed spiritual and temporal power firmly in Rome's hands. Emperors and kings heard these claims with growing unease.

Imperial advisors examined the language used in papal letters and court proceedings. They recognized how often the Donation was invoked to defend papal independence from royal oversight. Some clerics within the empire hesitated to press the claim too forcefully, wary of provoking retaliation. The text did not rest quietly in archives. It stood at the center of debate, cited in councils and contested in courts, as pressure mounted between throne and altar.

Scholars Weighing Words and Dates

In the fifteenth century, the humanist scholar Lorenzo Valla examined the Latin of the Donation closely. He compared its vocabulary and phrasing with known texts from Constantine's era. Valla noted terms and administrative concepts that did not exist in the fourth century. His analysis, written in 1440, argued that the language fit a much later period. This critique circulated among scholars, though it did not immediately alter church practice.

Earlier defenders of the Donation had relied on tradition rather than linguistic study. They copied the text into canon law collections and cited it in official letters. After Valla's critique, some church authorities limited discussion of his findings. Surviving records show that his work spread gradually through manuscript copies. The text remained in legal compilations, but scholarly scrutiny began to expose gaps between its claims and historical evidence.

A Scholar Under Suspicion

Lorenzo Valla's critique did not pass without cost. Though he worked under the protection of Alfonso V of Aragon, his challenge to a long accepted document drew sharp reaction. Supporters of papal authority questioned his motives. Some accused him of weakening the Church at a time of political strain. His careful study of grammar and vocabulary was seen by critics as an attack on sacred tradition.

Within clerical circles, those who read Valla's work faced their own pressure. Open agreement could bring isolation or loss of favor. Silence offered safety, but it left doubts unresolved. In university halls and private studies, scholars weighed loyalty against evidence. The parchment text still appeared in canon law collections, yet the men who examined its language carried the strain of challenging a claim that had shaped authority for centuries.

Preserving the Text Amid Debate

Following Valla's findings, church officials maintained the Donation of Constantine in manuscript collections, carefully cataloged and stored. Copies remained in archives of monasteries, cathedral chapters, and papal libraries. Access was limited to trusted clerics and scholars. No widespread alterations were recorded. The text continued to circulate in its original form despite emerging doubts about authenticity.

Over time, the document moved between institutions for safekeeping. Manuscripts were copied to preserve the content, with marginal notes added to clarify provenance or context. At some points, access required written permission from the head of a library or chapter. Letters and inventories document these procedures, showing careful tracking of the codices. The text was not destroyed or publicly challenged, yet officials exercised restraint, controlling its movement and study to balance archival preservation with sensitivity to emerging scrutiny.

Authority Framed in Ink and Doubt

In the decades after Valla's critique, the Donation of Constantine continued to influence disputes over papal and imperial authority. Rulers and clerics cited it in arguments over territorial control and investiture rights. Even as the Church grew in political power, the document remained a reference point for asserting claims, particularly in legal and diplomatic correspondence.

Yet uncertainty lingered. Scholars debated the text's origin, and some questioned whether relying on it could justify contested actions. The manuscripts were preserved, but their authority rested on tradition as much as historical truth. In councils and courts, officials invoked the document cautiously, aware that it might not withstand close scrutiny. Over time, its existence shaped practice and perception alike, leaving a legacy intertwined with both the assertion of papal power and the shadow of scholarly doubt.

Manuscripts Preserved Across Centuries

Several medieval copies of the Donation of Constantine survive today in European libraries, including the Vatican Library and the Biblioteca Medicea Laurenziana in Florence. The parchment shows slight discoloration and evidence of frequent handling, with fold marks and smudged ink along the edges. Marginal notes by later scholars and canon lawyers provide insight into how the text was studied and referenced over time. The documents are bound and protected behind glass, with controlled humidity and lighting to prevent further degradation.

Modern researchers examine the script, vocabulary, and codicology of each manuscript. Linguistic analysis confirms that the text belongs to the eighth century rather than the fourth. Conservators handle the parchment with gloves, recording every page digitally for study without repeated physical contact. Visitors can view the copies behind secure cases, reading transcriptions and translations. In these quiet halls, the parchment rests as a witness to both the claims of imperial grant and the enduring curiosity of scholars who sought to reconcile faith, law, and historical truth.

Chapter 17
The Tablet That Bound the People — The Aztec Tribute Stone and Imperial Obligation

Under a bright sky above Tenochtitlan, laborers drag a wide, carved stone across a sun-baked plaza. The city rises around them, temples and palaces reflecting in the waters of the surrounding lake. The slab is heavy, polished in parts, rough in others. Its surface is etched with symbols representing tribute goods and the towns obliged to deliver them. Each glyph is clear and precise, a visible mark of imperial command.

Officials stand nearby, marking the arrival of delegations from subject cities. They watch as bundles of cloth, baskets of maize, shields, and feathered ornaments are counted and aligned beside the carved images. The stone does not speak in words, but its carvings communicate obligation and authority. Merchants, warriors, and commoners pause to see the records, understanding that the tribute is not optional. The plaza is filled with the murmur of activity, the clatter of goods, and the careful attention of those enforcing the empire's reach.

As the laborers lift the stone into its final position, sunlight glints off carved surfaces. The inscriptions will endure, measuring compliance and marking the power of the Mexica long after the goods have been collected.

An Empire Measured in Stones

In the fifteenth century, the Aztec Empire governed a network of subject towns through a system of tribute that reinforced central authority. Tenochtitlan, the island capital, oversaw a vast region of lakes, causeways, and canals. Local lords remained in place, but their obligations to deliver specific goods were exacting and carefully recorded. Officials traveled to provinces to verify deliveries, ensuring that the capital received the materials needed for ceremonies, markets, and the ruler's household.

The tribute stones, like the one being installed in the plaza, were part of this administrative structure. Each carving displayed towns, goods, and quantities in a visual language understood by scribes and officials. Observers, from traders to warriors, recognized the authority embedded in these markings. The imperial bureaucracy relied on these durable records to maintain control, standardize obligations, and coordinate the flow of wealth across a diverse and populous region. Compliance was enforced publicly, with the plaza serving as a space where the empire's power could be measured at a glance.

Carving Authority in Stone

The tribute stone was crafted from a single slab of dense volcanic rock, selected for its durability and flat surface. Skilled artisans smoothed sections with stone tools before incising glyphs with obsidian and flint chisels. Each symbol represented a town, a type of tribute, or the quantity required. Lines were laid out in orderly rows, allowing officials and visiting delegations to read obligations at a glance.

Carvers worked under the supervision of imperial scribes, who verified that each depiction matched the official records of deliveries and obligations. Corrections were made before the stone left the workshop, ensuring accuracy. Once complete, the slab was transported to the central plaza on wooden rollers and ropes, where it was positioned upright for maximum visibility. The physical act of creating and installing the tablet tied labor, governance, and authority together, leaving a permanent record of the empire's reach and the responsibilities of its subjects.

Eyes on the Carved Record

When the stone was first installed, Aztec officials gathered in the plaza to inspect the carvings. Scribes compared each glyph to their own records, confirming towns, tribute types, and quantities. Nobles and tax collectors stepped forward, studying the inscriptions carefully. Their eyes traced the rows of maize, cloth, cacao, and feathers, ensuring nothing had been omitted or misrepresented. Small corrections were noted with paint, signaling minor adjustments without altering the stone's permanence.

Among the towns' envoys, hesitation moved quietly through the crowd. They recognized their obligations laid bare before the empire and the public. One miscount or misrepresentation could bring sanctions. Warriors and commoners observed the ritual of placement, sensing the weight of authority embodied in the carved symbols. Even before the first goods were delivered, the stone communicated both order and expectation, quietly imposing compliance through its presence in the heart of Tenochtitlan.

Tension Around the Plaza

As weeks passed, the stone drew attention from neighboring provinces and regional officials. Nobles compared its inscriptions to previous tribute records, noting variations in quantities or goods delivered. Some envoys worried that discrepancies might be interpreted as defiance. The carved tablet became more than a record. It was a measure of loyalty, a visible standard of compliance that could expose shortcomings.

Within Tenochtitlan, priests and imperial administrators monitored the stone closely. Certain glyphs marked offerings tied to religious ceremonies, heightening the stakes. Observers whispered about towns that had delayed deliveries or failed to meet quotas. The combination of public display and ritual significance increased pressure on all involved. Officials knew that mismanagement or misinterpretation could provoke conflict, yet every day the carved obligations remained unyielding, a silent reminder of authority and accountability.

Measuring Tribute and Enforcing Order

Scribes and officials repeatedly consulted the stone to verify deliveries from distant provinces. They cross-checked glyphs with account scrolls, marking any inconsistencies for review. Some towns were visited again to confirm quantities of maize, cloth, or feathers, ensuring the stone's inscriptions matched real tribute. Surviving colonial records note that repeated inspections helped standardize compliance and reduced disputes over obligations.

Later chroniclers suggested that the stone could punish towns automatically, but contemporary evidence shows a more measured process. Officials interpreted the carvings and determined whether sanctions or reminders were necessary. Access to the tablet was limited to authorized personnel, and only senior administrators could adjust counts in the records. The stone itself remained immobile, but oversight ensured that it functioned as a tool of governance rather than a symbolic warning. Its authority depended on careful study, verification, and controlled enforcement.

Burden Carried by Scribes and Envoys

Scribes tasked with recording tribute faced constant pressure. Errors in copying glyphs or misalignments in the stone could bring blame from imperial overseers. Some were reassigned or punished if discrepancies appeared between the carved tablet and provincial deliveries. Long hours bent over rough rock or inked scrolls left hands and eyes strained, yet accuracy was demanded above all else.

Envoys and local lords also bore the weight of public scrutiny. Their towns' contributions were permanently recorded, visible to rivals and visiting officials. A miscount or delayed delivery could damage reputation and standing within the region. Families, too, were affected, as failure to meet obligations could invite sanctions that touched household property or labor. The stone's permanence made every mark consequential, leaving those responsible for compliance under continuous strain and careful observation.

Procedures Etched in Stone

Over time, the Aztec administration developed formal protocols for handling the tribute stone. Officials recorded any changes in deliveries on scrolls, ensuring that what appeared on the plaza matched reality. When new provinces were brought under imperial control, scribes updated account rolls before transferring the corresponding glyphs to supplementary tablets or painted records. Access to the stone itself was restricted. Only senior administrators and selected priests could approach it directly during inspections or ceremonies.

During periods of unrest or military campaigning, the stone was protected by guards, and temporary markers were used to track incoming tribute until the full record could be confirmed. Written inventories documented each visit and adjustment, preserving continuity across years and rulers. Some towns were assigned officials who traveled with copies of the stone's inscriptions, reinforcing uniformity and oversight. Even in silence, the tablet remained central to the empire's administrative network, its fixed surface guiding enforcement and verification across a vast and complex system.

Legacy Carved in Stone

Years after its installation, the tribute stone continued to shape imperial administration. Officials referenced it when disputes arose over unpaid or miscounted goods. Provincial rulers adjusted deliveries to match what had been recorded, knowing the stone's markings could be cited in enforcement. Over time, it became a standard against which compliance was measured, giving long-term stability to the tribute system.

Yet uncertainty remained. Not all towns followed the tablet precisely, and some deliveries were disputed. Subsequent rulers occasionally added or removed officials tasked with monitoring the record, creating gaps in oversight. Even with careful administration, scribes noted that minor errors or ambiguities could stir tension. The stone's permanence preserved obligations, but human interpretation determined enforcement. Its presence anchored authority, yet the empire continued to grapple with the limits of control and the possibility that some towns might fall short despite the visible record.

A Stone Preserved for Centuries

The Aztec tribute stone, known today as the Stone of Tizoc or related provincial tablets, survives in museum collections such as the National Museum of Anthropology in Mexico City. Its carved surface remains legible, showing glyphs that record towns and their obligations. Conservators monitor humidity and light to protect the volcanic rock from further erosion, and the stone is displayed on a raised platform to prevent handling and accidental damage.

Scholars examine the inscriptions alongside surviving account scrolls and codices, comparing glyphs to reconstruct the empire's tribute network. Microscopic analysis confirms chisel marks and pigment traces used to highlight details. Visitors can view the tablet behind protective glass, observing the precision of lines and symbols etched centuries ago. The stone endures as a tangible witness to a system that bound people and places to the Aztec state, a silent record of labor, obligation, and imperial authority carved deep into the earth and preserved for generations to study and contemplate.

Chapter 18
The Sword That Crowned the Realm — The Japanese Imperial Regalia and Sacred Rule

In the quiet chamber of the imperial palace, polished wooden floors reflect the soft light of paper lanterns. Attendants kneel in measured rows, robes brushing the floor, hands folded. Outside, rain taps against tiled roofs, a gentle percussion that echoes through empty corridors. At the center of the room rests a lacquered chest draped in white silk. Inside lie three objects of immense significance: a curved sword, a polished mirror, and a string of jewel-like beads.

Priests and court officials move with careful attention, opening the chest and lifting each item with reverence. The sword glints faintly, its blade never drawn. The mirror rests in an ornate stand, reflecting candlelight but never facing the public. The jewel, smooth and dark, is threaded on fine silk. No one touches the objects without ceremony. The chamber is silent except for the rustle of fabric and measured murmurs of prayer. Each item carries authority, unseen yet absolute, confirming succession and binding the imperial line in ways that words alone could never achieve.

Power Rooted in Ritual and Lineage

During the early Yamato period, the Japanese state was consolidating authority across the archipelago. Regional clans pledged loyalty to the emerging imperial line, yet power was often contested through marriage alliances, military force, and religious sanction. The emperor's legitimacy depended not only on descent but on the recognition of ritual objects that embodied continuity and divine favor. These objects, sword, mirror, and jewel, served as tangible proof of authority, linking the ruler to Amaterasu, the sun goddess.

Court ceremonies reinforced this system. Only select priests and attendants were permitted to handle the regalia, and their training ensured that rites were performed precisely. Provincial leaders and military commanders understood that succession was validated through these objects rather than solely by proclamation. The combination of sacred mandate and controlled ritual created a structured hierarchy. Authority flowed through ceremonial possession, embedding the objects at the core of governance, while reinforcing loyalty and hierarchy across distant provinces.

Crafted for Heaven and Empire

The Japanese Imperial Regalia were assembled over centuries, with each object reflecting distinct materials and craftsmanship. The sword, Kusanagi, was forged from folded steel, polished to a mirror finish, and carefully curved for ceremonial use rather than combat. The mirror, Yata no Kagami, was cast in polished bronze, its surface smoothed to reflect light with clarity, while the jewel, Yasakani no Magatama, was carved from curved, translucent stone and threaded on fine silk.

Artisans worked under strict guidance, ensuring that each item conformed to ritual specifications. The sword's edge was never sharpened for battle, the mirror was never exposed publicly except in ceremonial contexts, and the jewel's form emphasized continuity and aesthetic perfection. Later records note careful transport and storage, with priests and attendants trained to handle, lift, and present each object. Together, the three items formed a tangible symbol of imperial authority, crafted with skill and intent to convey legitimacy, divine favor, and unbroken succession through centuries of ritual observance.

Eyes Turned to Sacred Objects

When the regalia were first displayed during succession ceremonies, court officials and provincial envoys watched with measured attention. The sword gleamed beneath candlelight, the mirror caught reflections from every angle, and the jewel rested on its silk thread. Attendants and priests noted each movement, ensuring that the objects were handled according to strict ritual protocol. Their presence confirmed legitimacy, yet required precise care.

Even within the court, uncertainty lingered. Junior officials hesitated to approach the chest, aware that a misstep could be interpreted as disrespect or incompetence. Regional leaders, observing from a distance, understood that possession of the regalia meant authority itself. The items were rarely seen outside ceremonial moments, yet their symbolism weighed heavily on those present. Each person acknowledged that the transfer of these objects, seen or unseen, would mark the continuation of power, even as the exact details of their handling remained tightly controlled and shrouded in ritual secrecy.

Anxiety Beneath the Courtyard Roofs

As succession approached, tension mounted among courtiers and regional leaders. Rival clans watched closely, aware that control of the regalia could determine claims to the throne. Priests responsible for ceremonial oversight debated the timing and order of presentation, knowing that any error might be seen as favoring one faction over another. The court remained silent, every movement measured, as whispers circulated about potential disputes.

Outside the inner chambers, provincial envoys and military commanders waited for confirmation of authority. The sword, mirror, and jewel were more than sacred objects. They were instruments that could legitimize or challenge rulership. Even experienced attendants experienced quiet apprehension, realizing that mishandling or misinterpretation could spark conflict. Each day the chest remained closed, each ritual carefully rehearsed, and each glance cast toward the regalia carried weight, creating a growing atmosphere of vigilance and subtle unease within the palace.

Guardians of Sacred Authority

Court records indicate that the sword, mirror, and jewel were kept under strict surveillance. Only appointed Shinto priests and trusted attendants were allowed to handle the regalia, and detailed instructions dictated how each item was to be lifted, displayed, and returned to the chest. Observers were barred from touching or even approaching the objects without express permission, ensuring that their presence, and the authority they conferred, remained controlled.

Later legends suggested mystical powers imbued in the regalia, but contemporary documents focus on practical oversight. Detailed logs describe the scheduling of ceremonies, the rotation of attendants, and the verification of each object's condition. In some instances, temporary relocations were documented when officials moved the regalia for palace renovations or in response to regional unrest. The careful management ensured that the regalia maintained both physical integrity and ceremonial authority, and that any claims to legitimacy rested on structured ritual rather than improvisation or unauthorized interpretation.

The Weight of Custody

Attendants and priests charged with handling the regalia carried a heavy burden. Any mistake, slight misplacement of the sword, improper positioning of the mirror, or tension in the jewel's silk thread, could be interpreted as negligence or disrespect. Court records note that even minor errors prompted private reprimands, and some attendants were reassigned when ceremonial standards were not met. Their vigilance was constant, and the strain of perfection weighed on both body and mind.

Regional leaders and junior officials also felt pressure. Observing the transfer and presentation of the sacred items, they understood that authority flowed through the objects, not merely the emperor's word. One misstep in protocol could call loyalty into question or give rival factions leverage. Families of attendants and priests shared the risk. Their social standing and opportunities at court were intertwined with the flawless execution of these duties. The regalia's presence made every action highly consequential, leaving those involved under continual scrutiny and quiet apprehension.

Rules Written Around the Regalia

Over generations, the imperial court formalized the handling of the sword, mirror, and jewel. Written instructions dictated who could touch each item, when and how they were to be moved, and which ceremonies required their presence. Temporary relocations for palace repairs or security concerns were recorded in official logs, detailing attendants, transport methods, and guarding procedures. Access remained tightly controlled, with only senior priests and appointed court officials permitted within the inner chambers.

Copies of procedural instructions were maintained in multiple archives to ensure continuity across reigns. During periods of political tension, administrators increased supervision, posting extra attendants and limiting the number of witnesses. Notes and memoranda described the care of the lacquered chest, proper wrapping of silk, and ceremonial placement of each object. Even in days without formal ritual, the regalia's position and preservation were documented, reinforcing continuity and stability. The detailed record of handling preserved both the objects and the authority they embodied.

Authority Bound in Objects

Over decades, the sword, mirror, and jewel shaped succession and legitimacy in Japan. When an emperor ascended the throne, possession of the regalia confirmed the transition, reducing disputes over lineage. Even during periods of political fragmentation, the items remained central to ceremonial recognition, reinforcing continuity of the imperial line across generations.

Yet uncertainty lingered. Rival clans occasionally questioned access or protocol, and some officials hesitated during ceremonies, aware that missteps could cast doubt on legitimacy. Records note that transport during conflict or palace renovation caused anxiety among attendants and observers alike. Though the objects endured physically, their authority depended on careful ritual and faithful execution. Each new transfer renewed both reverence and tension, leaving succession contingent not only on bloodline but on the presence and controlled handling of these sacred instruments.

Treasures Preserved in Silence

Today, the Japanese Imperial Regalia are held under strict security within the Imperial Palace in Tokyo. The sword, mirror, and jewel remain largely unseen, handled only by select priests and palace attendants during ceremonial occasions. Archival records meticulously document their custody, noting the names of those responsible for transport, display, and safeguarding. Climate controlled storage ensures the sword and mirror remain free from corrosion, while the jewel retains its smooth, translucent surface.

Scholars and historians rely on these records, rather than direct examination, to study the regalia. Written accounts of ceremonies, transport logs, and court chronicles provide detailed descriptions of their appearance, arrangement, and ritual use. Visitors rarely see the objects, yet the story of their role in confirming imperial succession endures. Even in quiet halls, the sword, mirror, and jewel convey authority, binding generations of emperors and attendants alike. Their physical presence continues to anchor the line of succession, a tangible reminder of power entrusted to carefully maintained ritual and centuries of tradition.

Chapter 19
The Tablet That Rewrote a Kingdom — The Sumerian King List and Political Memory

In a small scribal school in ancient Mesopotamia, a young apprentice presses a reed stylus into damp clay, feeling the cool surface yield beneath his fingers. The tablet rests across his lap, wide and rectangular, while his teacher kneels nearby, watching each stroke. Names of kings rise in careful columns, their reigns recorded in sequence, some impossibly long, others brief and troubled. The room smells faintly of wet earth and reed mats, and sunlight filters through a high window, casting shifting lines across the students' work.

Other apprentices lean forward, reeds poised, observing as the list takes shape. The sequence of cities and rulers moves from Uruk to Kish, Lagash to Ur, marking the transfer of authority through the ages. The tablet is not merely a record. It is a claim about who rules, when, and where. Each line carries weight beyond the clay. Outside the school, the city hums with activity, but within these walls, the apprentice inscribes the memory of kings, shaping how generations will understand the passage of authority in the land of Sumer.

Cities Competing for Legacy

In early second millennium BCE Mesopotamia, city states such as Ur, Uruk, Kish, and Lagash jostled for dominance. Kings ruled through armies, temple authority, and trade alliances, yet no single ruler commanded the entire region. Authority was fragmented, and succession often sparked disputes among rival families or factions. Temples, which controlled economic and religious life, recorded transactions and appointments, creating administrative frameworks that guided governance.

Within this network, scribes maintained lists of rulers to preserve memory and legitimize current power. Royal and temple officials used tablets to document succession, linking kingship to cities and divine favor. These records were copied, preserved, and referenced when disputes arose or new leaders took the throne. The practice gave emerging rulers a tangible connection to past authority while reinforcing institutional hierarchies across city states.

Clay tablets became central to governance, providing both a practical and symbolic measure of continuity in a landscape of shifting power.

Clay and Cuneiform Craft

The Sumerian King List was inscribed on clay tablets, prepared by kneading and smoothing local riverine clay into flat surfaces suitable for writing. Apprentices and master scribes used reed styluses to press wedge shaped cuneiform symbols into the damp material. Each column recorded a ruler's name, city, and the length of reign, with legendary early kings stretching across implausible spans and later rulers marked with more realistic lifetimes.

Creation required careful attention. Scribes worked from prior copies or oral accounts, verifying sequences where possible. Tablets were sometimes baked to preserve inscriptions, ensuring they survived both handling and environmental exposure. Multiple versions existed, with small variations between copies, reflecting the needs of different city states or temple archives. Each tablet functioned as both a practical record for contemporary officials and a lasting monument to political memory, providing successive generations with a structured account of rulership and succession in early Mesopotamia.

Reading Authority in Clay

When the King List tablets were first inscribed, temple officials and scribes examined them closely. Senior priests traced the columns of names with careful fingers, confirming the sequence of rulers and the cities associated with each reign. Apprentices watched and copied the symbols, learning both cuneiform and the historical memory of past kings. The tablets were handled with care, recognizing that errors could misrepresent authority or upset administrative order.

Even among experienced scribes, hesitation lingered. Some names stretched back into legendary reigns, with lifespans that seemed impossible. Officials whispered among themselves, questioning how to reconcile these long ago rulers with known history. Despite the uncertainties, the tablets were accepted as guides for current and future administration. Their presence in the temple archives and schools lent weight to succession claims, while quietly reminding all who approached them that the past, recorded in clay, carried both authority and the responsibility of careful attention.

Anxiety Across the City States

As the King List circulated among temples and royal archives, tension grew among city officials and scribes. Each tablet recorded not only succession but the legitimacy of current rulers. If a name or reign was disputed, questions arose about whether the current king's claim rested on accepted precedent. Priests debated quietly in chambers, concerned that errors could undermine ceremonial authority or provoke rival cities.

Regional rulers and temple administrators watched carefully. The order of kings linked cities to past victories, divine favor, and political hierarchy. Any inconsistency could embolden challengers or spark local unrest. Scribes and officials felt the pressure of precision, knowing their work was being scrutinized by both peers and higher authorities. The clay tablets, silent yet immutable, became a measure of political legitimacy, and the responsibility of maintaining their accuracy weighed heavily across generations of bureaucrats and temple staff.

Tablets Examined and Copied

Scribes and temple officials worked to ensure the King List reflected accepted succession. Copies were made for different archives, each carefully compared against earlier tablets. Officials sometimes annotated minor discrepancies, noting differences in reign length or city affiliation. Surviving fragments show these efforts to maintain accuracy, with corrections impressed into clay or added along the edges.

Later historians claimed the list was purely legendary in its early sections, but contemporary records indicate that officials treated it as a practical tool for governance. Access was limited to trained scribes and priests, and only senior administrators could authorize new copies or revisions. The tablets were never freely circulated among the populace, and their placement within temples reinforced both ceremonial and administrative authority. By controlling who could study and copy the text, the city states preserved both its physical integrity and its function as a measure of legitimacy.

The Burden of Accuracy

Scribes responsible for the King List carried significant pressure. Each symbol, each mark of a ruler's name, could influence the perceived legitimacy of kings across city states. Mistakes in copying or recording reign lengths could draw criticism from priests or officials, and in some cases jeopardize a scribe's position. Apprentices learned early that a single error might undermine both reputation and career prospects, leaving them anxious under the watchful eyes of senior masters.

Temple officials and city administrators faced their own strain. The tablets influenced succession disputes and territorial claims, so any perceived misalignment of records could trigger tension between rival cities. Families of scribes and officials shared the stakes. Social standing and access to positions depended on flawless performance. Quiet isolation often followed mistakes, as peers distanced themselves from those whose errors might reflect badly on the temple or city administration. The tablets' permanence made every action consequential, leaving those entrusted with them under constant scrutiny and enduring personal pressure.

Preservation Across Generations

Over time, city states established formal procedures for maintaining the King List. Tablets were stored in temple archives under the supervision of senior priests and officials. When new rulers ascended or city boundaries shifted, copies were made and cross checked against older tablets. Access was restricted, only trained scribes could handle the clay, and all corrections were documented along margins or on separate fragments to preserve the original inscriptions.

During periods of conflict or relocation, tablets were transported carefully, wrapped in cloth or placed in protective containers. Records of each transfer were kept in accompanying account rolls. Administrative logs noted who moved the tablets, when, and under what conditions. Even in silence, the tablets were closely monitored, with procedural oversight ensuring that each copy accurately reflected the lineage of kings. This system reinforced continuity, allowing successive generations of scribes and officials to uphold the record as a central instrument of legitimacy and institutional memory.

Authority Preserved Through Clay

Decades after its creation, the Sumerian King List continued to shape perceptions of rulership across Mesopotamia. Officials consulted it when recording succession, resolving disputes, or asserting legitimacy over contested cities. By providing a continuous sequence of kings, the tablets reinforced the idea of orderly transfer of authority, even as individual city states rose and fell.

Yet uncertainty remained. Some reigns were legendary, with implausible lifespans, leaving scribes and officials to question how closely these records reflected reality. Gaps between versions and minor discrepancies created moments of hesitation during disputes. Even as tablets were carefully copied and preserved, their authority depended on interpretation and acceptance by successive rulers and priests. The clay itself endured, but its power rested on human judgment, ensuring that the memory of kings remained influential yet never entirely free from doubt or challenge.

Clay Tablets Preserved Through Millennia

Several copies of the Sumerian King List survive today in museums and archives, including the British Museum, where the Weld Blundell Prism is carefully displayed. The clay tablets bear wedge shaped cuneiform, some sections worn from handling or age. Conservators stabilize fragments, controlling temperature and humidity to prevent further erosion.

The tablets are supported on custom mounts to avoid stress, allowing scholars to examine inscriptions without damaging the ancient clay.

Modern researchers study the text alongside archaeological findings, inscriptions, and historical records to reconstruct political continuity in early Mesopotamia. High resolution imaging reveals details of line placement, stroke depth, and spacing that were once obscured. Visitors can observe the tablets behind protective glass, noting how careful arrangement and craftsmanship preserved centuries of administrative memory. Each prism and fragment embodies the Sumerians' efforts to record succession, offering a tangible link to rulers long past, and a quiet witness to how authority and memory were carved, preserved, and interpreted over millennia.

Chapter 20
Epilogue:
The Object That Outlived the Throne — Power Preserved in Material Form

After the last ruler touched a seal, a crown, or a carved tablet, the objects often remained in place, unobserved by ordinary citizens. In palaces, temples, and administrative halls, clerics and officials recorded inventories, noting the location and condition of each item. Guards were assigned to chambers, and access was limited to a select few, often those sworn to oath and secrecy. Records describe keys handed only to senior officers and strict procedures for unlocking storage rooms, emphasizing process over publicity.

Even when objects were no longer actively used, institutions continued to monitor them. Ledgers and scrolls documented their presence, noting repairs, environmental conditions, and ceremonial readiness. In some cases, rulers ordered items buried temporarily or hidden in fortified repositories to protect them from conflict. The physical traces of these artifacts remained, yet their public function ended. Silence became a form of control, allowing authority to persist even when objects were out of sight and out of use.

Hesitation in the Shadow of Knowledge

When knowledge embedded in objects was deemed dangerous, hesitation often preceded any action. Priests, scribes, and administrators debated whether texts or regalia should be displayed, copied, or confined. In Mesopotamia, certain clay tablets recording royal succession were removed from active administration when dynasties changed and stored in temple vaults instead of being destroyed. Similarly, in medieval Europe, documents like the Donation of Constantine were preserved in monastic libraries even after scholars questioned their authenticity. Marginal notes and restricted access reflected a balance between caution and preservation.

Officials frequently deferred decisions. A council might record the existence of a text or artifact but leave it untouched for decades, concerned that disclosure could inflame disputes or destabilize authority. In Japan, the Imperial Regalia were maintained under strict ritual oversight, yet largely removed from daily governance. The sword, mirror, and jewel remained unseen, their power preserved by ceremony, not action. The choices to retain, restrict, and silently observe created long-term uncertainty, leaving future custodians to navigate inherited responsibility without precedent.

Layered Records and Constrained Access

Archival traces provide insight into these patterns of restraint. Ledgers, inventories, and palace chronicles frequently document items long after their active use ended. Boundary stones in Mesopotamia, for example, were cataloged in temple registers even when no longer consulted in legal disputes. In Aztec administrative practice, tribute stones were preserved in temples after collection, their inscriptions studied periodically to verify accuracy, but access was controlled. These layers of record created redundancy, ensuring that knowledge survived even when the physical objects were no longer deployed.

Institutions established hierarchical access as a measure of safety. Only authorized scribes, priests, or officials could handle, view, or copy materials. Transport and handling were meticulously documented, with instructions for protective wrappings, locks, and ceremonial procedures. In Europe, copies of potentially contentious documents circulated only among trusted clergy. In East Asia, regalia were moved sparingly and only under careful supervision. The combination of restricted movement, careful copying, and archival oversight allowed objects and knowledge to endure while minimizing risk to institutional authority.

Long-Term Consequences of Containment

Even after active use ended, the presence of these objects influenced governance, ritual, and scholarly practice. In Mesopotamia, tablets recording royal succession continued to shape legitimacy debates, even when they no longer guided administrative decisions. In medieval Europe, preserved documents were cited during legal disputes, their authority reinforced by physical permanence and the prestige of containment. In Japan, the Imperial Regalia defined succession rituals for centuries, their unseen presence exerting influence over ceremony, loyalty, and hierarchy.

Hesitation and containment created tension that lasted for generations. Officials, scholars, and clerics were forced to balance access with control, using silence, omission, and restriction as tools of governance. In some cases, failure to act decisively was itself a policy, leaving objects in limbo while their authority persisted. The enduring presence of these items, combined with the deliberate restraint of institutions, illustrates how material artifacts outlasted the immediacy of power while shaping the behaviors and decisions of those who remained responsible for them.

The Persistence of Influence

Over centuries, objects once central to decision making became symbols and references for subsequent generations. Royal seals, crowns, and tablets were cited, displayed sparingly, and studied selectively. Inscriptions and carvings provided evidence of past authority, guiding interpretation of current governance without direct application. Scholars copied texts to preserve form and content, while custodians maintained physical integrity through careful handling and environmental control.

Instances of fear driven preservation recur across cultures. Scribes in Mesopotamia recorded disputed sequences of rulers to avoid political conflict. Papal archivists in Europe preserved contested documents to safeguard legitimacy while limiting access. In Japan, regalia were retained unseen, reinforcing succession while minimizing risk. Across all examples, the objects carried latent power. They were never inert, even when stored or hidden. Custodians understood that authority could be exercised simply through the existence of the artifact, a presence that compelled acknowledgment without active display.

Surviving Witnesses

The physical remnants of these systems offer testimony across millennia. Clay tablets in museums, preserved parchments, and carefully maintained regalia bear marks of handling, repair, and ritual. Inscriptions remain legible, seals show impressions, and careful mounts protect fragile objects from decay. Archival records document custodians, transport, and ceremonial use. Even when original intent was forgotten, the physical objects endured, carrying evidence of the systems they once enforced.

These surviving witnesses demonstrate continuity in institutional behavior. Decisions to restrict, guard, or quietly observe were documented alongside the objects themselves. Layers of record and physical preservation ensured that knowledge persisted while human actors changed. Later observers and scholars could study the evidence, but access remained limited, reflecting the cautious approach of the original custodians. The objects outlasted generations, carrying authority quietly but unmistakably.

Waiting in Silence

Over centuries, artifacts passed from active instruments of governance to objects of preservation. Museums, libraries, and treasuries now hold many of these items, maintaining environmental conditions, security, and protocols for handling. Researchers study them carefully, guided by historical records, but they are rarely moved beyond controlled access. The authority embedded in their material form continues to influence perceptions of legitimacy, succession, and institutional memory.

Even today, the lingering tension of control remains. Clay, stone, metal, and jewel bear silent witness to power, knowledge, and ritual. Custodians, then and now, navigate uncertainty, aware that the potential consequences of misuse extend beyond immediate observation. The objects endure, preserved against time, conflict, and human error. Their authority has not vanished; it waits, embedded in the material, poised to shape interpretation and action whenever attention returns.

Made in the USA
Monee, IL
04 November 2026

67381283R00098